Writing and Presenting Scientific Papers

Birgitta Malmfors

Phil Garnsworthy

Michael Grossman

Nottingham University Press
Manor Farm, Main Street, Thrumpton
Nottingham NG11 0AX, United Kingdom

NOTTINGHAM

First published 2000

Reprinted 2000, 2001, 2002

British Library Cataloguing in Publication Data
Writing and Presenting Scientific Papers:
I Malmfors, B., II Garnsworthy, P.C., III Grossman, M.

ISBN 1-897676-92-1

Typeset by Nottingham University Press, Nottingham
Printed and bound by The Cromwell Press, Trowbridge

CONTENTS

THE AUTHORS

Birgitta Malmfors is an Associate Professor in Animal Breeding and Genetics at the Swedish University of Agricultural Sciences, Uppsala, Sweden. She has written many scientific papers, reports and articles. She also has authored textbook chapters in animal breeding and a Swedish handbook on writing and presenting scientific papers. She regularly makes presentations at national and international meetings. She has taught students successfully for thirty years, and has been given a distinguished award for teaching and communication skills.

Phil Garnsworthy is a Senior Lecturer in Animal Production at the University of Nottingham, UK. He has edited 14 books of conference proceedings and contributed chapters to several other books. He also has written many papers for scientific journals and articles for radio, television and the popular press. He regularly makes oral and poster presentations at international scientific conferences and industry meetings. He has been teaching students for more than twenty years and is particularly interested in training students to communicate effectively.

Michael Grossman is a Professor in the Departments of Animal Sciences and Statistics at the University of Illinois, USA. He has written numerous peer-reviewed scientific papers as well as popular science articles, and has given national and international scientific presentations. He has been teaching in the areas of genetics and mathematical modelling for more than thirty years. He also has taught workshop courses on techniques for scientific writing to Ph.D. students at Wageningen University, The Netherlands.

PREFACE

Think of any great scientists in history and consider what made them great. You will probably say that they made a new discovery, invented something or interpreted existing knowledge in a new way. How do we know they did this? They communicated their findings to others! Without communication, science would not have developed. Even today, scientists are judged by their ability to communicate their ideas and findings through written papers and conference presentations. You can perform the most elegant research in the world, but it has no value unless you tell somebody the results.

Preparing papers and presentations can take quite a long time, but usually not as long as the time spent doing the research. Publication and presentation should be seen as an integral part of the research process – it is the only way to produce something that people will remember. Scientific communication requires a special language, so that your message is clearly and accurately conveyed to the reader or listener; its effectiveness can be enhanced by a few simple techniques. Young scientists and students can be trained in these techniques, but experienced scientists can also benefit from critically examining the way they write and present papers.

This book is dedicated to Professor Rommert Politiek, The Netherlands, former Editor-in Chief of the Elsevier journal "Livestock Production Science". He identified the need for improvements in many papers submitted, as well as in oral and poster presentations at scientific meetings. He initiated, with great enthusiasm, a workshop for the European Association for Animal Production (EAAP), to train young scientists in writing and presentation skills. The first workshop took place at the annual meeting of EAAP in Warsaw (1998). The workshop was so successful that it was repeated in Zurich (1999) and again in The Hague (2000). The first two workshops were sponsored by EAAP and the third jointly by Elsevier and EAAP. We three authors of this book were encouraged by Professor Politiek to be the main tutors at these workshops. Having produced a booklet for the first workshop, we decided to gather our ideas together in the form of a book that would reach a wider audience.

Other books have been published on various aspects of writing and presenting scientific papers, some covering just one facet, others going into much more depth; a number of these books are included in our listing of literature for further reading. Our book is intended to be Accurate and Audience-adapted, Brief and Clear (the communication ABC). It covers the whole "chain" of science communication, including literature searching, writing for journals, conferences and the popular press, presenting papers orally or as posters, and training students in communication.

The book is aimed at scientists of all ages, university students and university teachers – anybody involved in communicating research results, or in training students. We particularly hope that young scientists will find the book useful and that it will encourage them in their communications. Although the EAAP workshops were designed for Animal Scientists, the information has been adapted to be useful for scientists and students from all disciplines.

Producing this book has been an interesting and valuable experience for us. We have learnt techniques from each other and we have broadened our perspectives. Although we all speak "English" (native and non-native), our origins are Sweden, UK and USA. This mix of writing cultures has emphasised to us that there is usually no "correct" way to write or present. For the convenience of our UK publisher, the book is written mostly in "British" English, but we have always tried to remember the need to keep the message clear for an international audience.

All drafts were produced directly on the computer, using ordinary software for word processing, graphics and presentation. Many of the illustrations were produced using Clip Art from Microsoft under the terms of their end-user agreement, which allows licensed users to use clip art in any publication, providing the publication itself is not a collection of clip art. We found electronic mail invaluable for the exchange of documents and ideas, which allowed us to work from three different countries.

We also received very valuable help while working on the manuscript; special thanks are addressed to Professor Jan Philipsson, Sweden, for reading drafts and providing useful suggestions, to Sarah Keeling, UK, for typesetting the manuscript, and to Nancy Boston, UK, for reading the final proofs.

Do not expect to find "rules" in this book; it's the journals, publishers and conference organisers that set rules for you to follow. Each of us writes and presents in different styles, so we do not have any definitive answers. What we do provide, however, are guidelines for good practice, based on personal experience and observation. We hope that you will find these guidelines useful in your scientific communications.

Birgitta Malmfors
Phil Garnsworthy
Michael Grossman

July 2000

1

COMMUNICATING SCIENCE

The aim of research is to contribute to knowledge, so that every new result adds to the previous state of knowledge, forming a basis for new thinking and interpretation, new implications, identification of needs for further research etc. Research results, however, do not contribute to knowledge and development unless they are communicated effectively. Effective communication of science, both to scientists and to other audiences, is a very important component of the research process.

Communication is needed, not only to spread research results, but also to formulate results. Writing or talking about your research helps to clarify your thoughts, and to put your research into a deeper and wider context. Therefore, start writing and talking about your research early in the process. Communication is a vital part of your research!

Scientific communication occurs in many forms, such as papers in scientific journals, reports, conference papers and abstracts, graduate- and post-graduate theses, review papers, proposals, popular science articles, newspaper articles, computer-mediated information, oral presentations, posters, interviews and discussions. Various forms of communication have a great deal in common, but they also differ, e.g. in purpose and audience addressed. They also differ with regard to scrutiny before publication/presentation.

A good knowledge on how to handle various forms of scientific communication is needed in many professions, not only for scientists. It is therefore essential to include the topic at every level of education.

THE ABC OF SCIENCE COMMUNICATION

Communicating science usually means communicating *new* knowledge or summarising the present state of knowledge. It is important for the audience to catch the message with as little misunderstanding as possible and to feel confidence in what is written or said.

The ABC of science communication is that it should be:

- **A**ccurate and **A**udience-adapted
- **B**rief
- **C**lear.

Science is international. This means that many of those who read or listen to a scientific presentation will be doing so in a foreign language. This further emphasises the need for clarity and for the presentation to be logical, consistent and coherent. Communication is a two-way process. Information cannot merely be delivered – it must be received and understood as well. The message delivered may be accurate, brief and clear, but yet not be received and understood. This may happen if what you write or say does not relate to the frames of reference of the audience. Adapting to the audience, therefore, is very important.

A basis for the scientific process is to formulate a hypothesis, which means that you pose a question and a hypothetical answer. Questions and answers are the basis for communication as well. For effective communication you cannot just think of your own topic and the message you want to deliver. You must also consider what questions your audience might have with regard to your topic. Some components of effective communication are indicated in Figure 1:1.

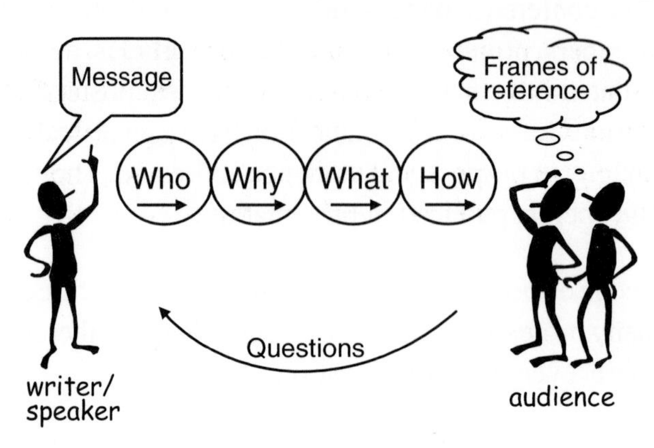

Figure 1:1. Some components of effective communication.

When preparing to write a paper or a report, or to make a poster or an oral presentation, start by asking yourself the questions: Who? – Why? – What? – How?

- **Who** are you addressing: scientists who are specialists in your field of research, a wider group of scientists, fellow students, or public audiences?
- **Why** is your message important? Why are you communicating it? Presumably not just for merits, but to add to the knowledge pool, to teach, to inform, to persuade and push for development.
- **What** are your main findings or "take-home" messages? What are you going to present – new research results or a review of a topic? What prior knowledge, expectations and questions may your audience have with regard to your topic? What technical language do they understand?
- **How** can you best deliver your message and satisfy the audience's needs? How will the audience use their new knowledge?

The order of the words, who-why-what-how, and their relative importance may vary in different forms of scientific communication. This is also true for the emphasis given to your own message and to the anticipated audience questions. In a scientific paper you might concentrate on research results, whereas when writing a popular article or giving a talk, the audience questions are more accentuated. Nevertheless, you should always adapt to the audience's prior knowledge in any type of communication.

SCIENTIFIC VERSUS POPULAR SCIENCE WRITING

Scientists usually communicate the same topic in various ways and to different audiences. New research results are often first communicated to other scientists at a conference, both in written form and as an oral presentation or as a poster. The written conference paper (sometimes just an abstract) normally follows the rules of scientific writing, but is generally not reviewed by peer scientists before delivery.

The core communication of new research results is a paper published in scientific journals. To meet the specific demands on clarity in communication of new original knowledge, the scientific paper is organised in a standard way. The format normally used is called IMRAD, referring to the main sections of the paper: Introduction, Materials and Methods, Results and Discussion. The common practice for journals is to have the scientific paper reviewed by peer scientists anonymous to the author before it is accepted for publication. This is done to ensure that the results and interpretations published are of good quality.

New findings in research need to be communicated also to public audiences, which means that popular science publishing and presentation are required in addition to the scientific communication. Some features of scientific and popular science writing are given in Figure 1:2.

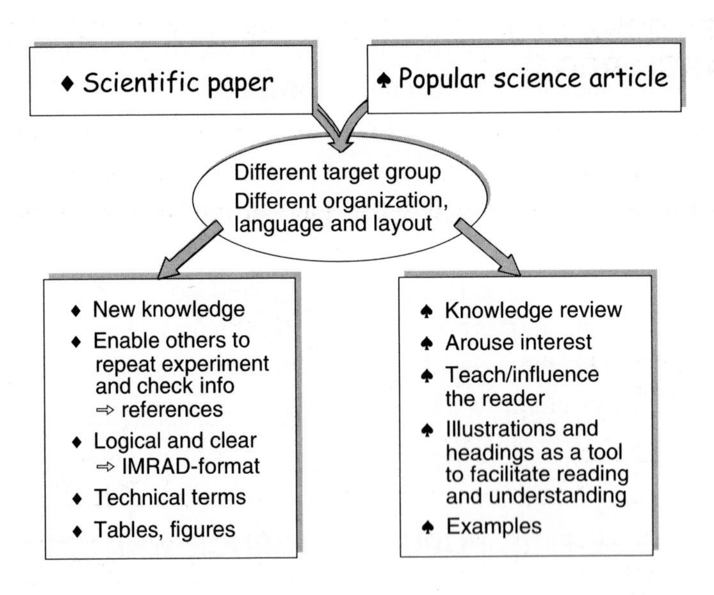

Figure 1:2. Some characteristics of scientific and popular science writing.

Scientific and popular science writing differ in several aspects, but there are also many similarities. The communication ABC, i.e. accuracy, brevity and clarity, should be fulfilled in each type of writing, but of course the technical language cannot be the same for non-specialists as for specialists. The specific demands on scientific papers originate from the fact that original documentation of new research results requires precision. This precision includes distinguishing the new results presented in the paper from previous research results, and always giving the reference to the original source. Scientific writing should be logical and clear so that the reader understands what is written with minimal misinterpretation, and so that others can repeat or check what is done or said. In popular science writing, you first need to think about what in your research area might interest the reader, and then about how to explain things so they are understood even without previous knowledge in this area.

Whether writing a scientific or a popular science paper, it might be fruitful to have some characteristics of the other type in mind. Awakening interest and helping the reader is important also when writing a scientific paper. Furthermore, reliable information ought to be a guiding star not only in scientific but also in popular science writing.

THE SECTIONS OF A SCIENTIFIC PAPER REFLECT THE RESEARCH PROCESS

To better understand the organisation of a scientific paper or report, it is worthwhile looking at the various steps in the research process. These steps, as well as the corresponding sections of the scientific paper, are illustrated in Figure 1:3.

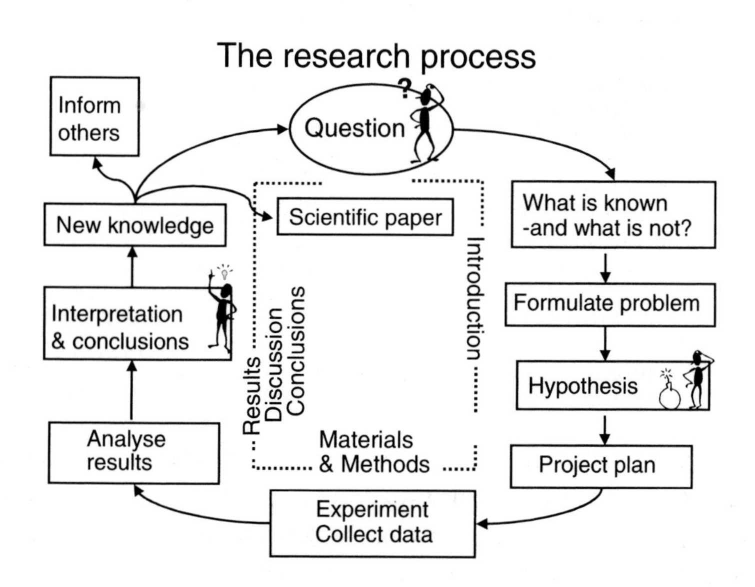

Figure 1:3. Major steps in the research process (outer circle) and corresponding sections of a scientific paper (inner circle). [1]

[1] After an idea of Backman, J. 1985. Att skriva och läsa vetenskapliga rapporter [Writing and reading scientific reports], p.17. Studentlitteratur, Lund.

Research is exciting! We deal with something unknown, a search for an answer initiated by a question. We want to know why things happen, why there are differences between treatments, how things can be explained. It is like looking for a piece to add to a puzzle.

An important initial step in the research process is to find out what is already known about the topic by reading the scientific literature, and maybe also by using other sources of information. Having done so, it is possible to identify what is known and what is not known, and to formulate a specific problem to study. The research task will be determined by your hypothesis, which must be testable and formulated in such a way that you can either reject or accept it, depending on your results.

To test the hypothesis it is necessary to collect data, e.g. make observations by doing an experiment, using field data, interviewing people. This requires a plan or experimental design, so that we know which materials and methods to use, and how the study should be structured to avoid confounding of different effects in the interpretation of results. The results of the study are analysed, usually by statistical methods, and then interpreted and discussed in view of the previous knowledge, the hypothesis, and the formulated problem. Conclusions are drawn and a new piece is added to the knowledge puzzle. The piece might make the picture more complete, or help to revise it. The outcome could be new recommendations and implementation, as well as identification of questions for future research. The research process thus forms a circle, where for each cycle the steps need to be documented and communicated to contribute to knowledge. The basis for this documentation is the scientific paper, which might also be supplemented with other forms of communication.

The scientific paper (the inner circle in Figure 1:3) reflects the research process, both in organisation and content. In the *Introduction* section you answer questions such as: “Why is the topic important, how does it relate to previous knowledge, and what was your hypothesis and objective?”

The *Materials and Methods* section gives details on how you did the study, i.e. the project plan or experimental design, the materials used, the methods for making observations, and the analysis of the data. The *Results* section describes what answers you obtained, often presented in the form of tables and figures. Results are discussed and interpreted in relation to the previous knowledge, the formulated problem and your hypothesis, either in a separate *Discussion* section, or in a joint *Results and Discussion* section. Your conclusions can be included in the Discussion part, or given their own heading: *Conclusions.* The scientific paper will also include an *Abstract* (an independent summary of your paper) and a list of *References*, so that others will know where to find the literature to which you refer in your text.

Producing the scientific paper is an integral part of the research process. It is obvious, therefore, that thinking of, planning and getting started in writing the paper can, and should, be initiated early in the process, especially because it helps to clarify your thoughts. Be aware, however, that writing takes time, and you will often have a deadline for delivery, especially if you are a student performing graduate or post-graduate research. So, document continuously what you do, make notes of your ideas, organise the results you obtain in tables and figures, and set up a good reference system to keep track of the literature used. Don't wait for all results and analyses to be completed before you start writing the paper or report. You will find hints on how to deal with this topic in the chapter "Getting Started in Writing", but first we will discuss the sections of a scientific paper in more detail.

2

SECTIONS OF A SCIENTIFIC PAPER

As discussed in the previous chapter, you will be asked to prepare various types of written communication during your professional career. Communication of science is as important to the scientific process as designing, conducting and analysing the experiment itself.

Journals publish different types of scientific paper, including comprehensive, full-length research papers; review papers; symposium papers; invited papers; technical research notes, which give results of complete but limited experiments; rapid communications, which deal with a "hot topic"; book reviews; and letters to the Editor, whose purpose is to discuss, critique, or expand on scientific points made in recently published papers.

Most scientific papers submitted to journals follow a standard format: Title; Abstract; Introduction; Materials and Methods; Results; Discussion; Conclusion; References. Acknowledgements and Appendicies are optional, as are tables and figures. Sometimes a section may be replaced by another (e.g. Theory may replace Materials and Methods in a paper that has a theoretical development) or even omitted (e.g. Materials and Methods, or Results and Discussion, may be omitted in a review of the literature).

Major headings for review papers, symposia papers, or short communications sometimes deviate from standard format. You might, for example, use section headings that are appropriate to the subject being reviewed. For all types of paper, follow the guidelines set by the journal or publisher to which you will submit the paper; journals usually publish these guidelines annually and refer to them in each issue. The contents of standard major headings are discussed in the following sections.

MAJOR HEADINGS

Title

The title should attract people to your paper

The title tells you what the paper is about, with the main purpose of encouraging people to read the paper. The title will be read more than any other section of the paper, so it should be informative, specific and concise. Make the title informative by describing the subject of the research, not results of the research. Put the most important words first. Make the title specific by differentiating your research from others on the subject, for example, by describing the novel materials used. Limit the title to probably not more than seven to ten words. Journals give instructions to authors as to the length of the title. Write a short working title first, but keep in mind that it may change as the focus of your paper changes while you write. Make sure that the final title is relevant to the content of the paper.

Only use technical terms if they are familiar to most readers and do not use abbreviations in the title. Eliminate "waste words", words that say nothing, e.g. "Observations of…", "Studies of…", "Investigations of…", "A Note on…", or "Examination of…". Avoid titles with a series, e.g. I, II, III, …. Besides appearing self-aggrandising, which you should avoid, you might publish the first paper (I) but never publish another in the series (many journals insist that all papers in a series are submitted together). For review papers, consider using a "hanging" title, e.g. Xxxxx: a review.

If you are writing a title to accompany an abstract for a presentation at a scientific meeting, remember that the title is often the primary basis for placing an abstract in a certain session of the scientific programme. Vague or uninformative titles increase the risk that the abstract might be allocated to an inappropriate session or even be rejected. Beware of long titles that can look awkward on a title slide or poster.

The *running head*, which has a maximum number of characters plus spaces, is an abbreviated title and appears as a header on pages of the journal. Sometimes the journal specifies what the running head must be, depending on your paper. If, for example, you submit a short communication, the journal might specify the running head (e.g. SHORT COMMUNICATION:), followed by a short title. Make sure that your running head conveys the essential message from the title, particularly in a review paper for a book; it should attract people as they browse through the pages.

Abstract

The abstract is especially important because a large number of people will read it (e.g. through abstracting journals), and often it will determine whether they read the entire paper. Readers should be able to understand the abstract without having to be familiar with the details of the research.

The Abstract can be read on its own

Whether you are writing an abstract for a paper or for a presentation at a scientific meeting, the abstract should describe the problem and summarise the major points of the research in a brief and understandable form. It should start with clear statements of the objective, the approach, and main results, and it should end with one or two sentences that emphasise important conclusions. Be specific and concise because the length of the abstract is often restricted to a maximum number of words or number of characters plus spaces, depending on the journal. The abstract should stand alone, so state results as facts but do not cite references to literature, tables, or figures. Avoid using an abbreviation, but if you must, consult the journal to be sure that it appears in the journal's list of standard abbreviations, which do not require definition (e.g. DNA for deoxyribonucleic acid).

The abstract, like other sections of the paper, is normally written in the past tense, but sometimes the present tense is used. Use the same tense throughout the section, or at least throughout a paragraph within the section. Changing tense within a section gives the appearance of uncertainty and inconsistency.

At the end of the abstract, list *key words* that best describe your research. Key words are used by indexing services and to form the subject index of the journal. Key words, therefore, should include when applicable the species, variables tested, and the major response criteria. If the key-word index is not based on the title, appropriate words from the title should be used as key words (see chapter "Literature Searching and Referencing").

Introduction

The introduction persuades the reader that the topic is important and that the objective of the research is justified. It motivates and justifies the research, by providing the necessary background, by explaining the rationale for the study, and by stating clearly the objective and the approach. The introduction should orient the reader by summarising **briefly** the relevant literature. Be specific and direct, but give the reader all necessary information leading logically to, and focusing quickly on, the objective and the approach. The introduction may also specify the question to be answered or the hypothesis to be tested.

Clearly state the objectives

Some journals want an extensive discussion of relevant literature to be in the discussion of results, not in the introduction. If you are writing a review paper, in fact, the literature is reviewed not in the introduction but in the discussion. Review papers normally have a short introduction, a long discussion of literature (split into topics), followed by conclusions; there may be a general discussion before the conclusions.

Materials and Methods

The materials and methods provide a clear and complete description for all experimental, analytical, and statistical procedures. Organise the section logically, perhaps chronologically, and use specific, informative language. Include necessary information, and omit unnecessary information. It is not necessary, for example, to describe in detail a procedure already published, so cite the original reference but explain modifications. This section should include enough information so that another researcher can repeat the procedures and expect to get the same results.

The design of planned experiments, or the sampling protocol of surveys, should be conveyed clearly and concisely. In particular, the replication and blocking structure should be described to allow the reader to reconstruct the experimental layout. Describe clearly and fully your experimental subjects, numbers, treatments, environmental conditions, measurements and statistical models. State any assumptions you made, and state quantities in standard units. Microorganisms should be named by genus and species; specific strain designations and numbers should be used when appropriate. If you mention an enzyme, then you should include the EC number. For drugs, chemical names are more universally descriptive than trade names. For surveys, describe the sampling procedures and observational methods.

Consult guidelines for contributors to the journal if you refer to sources of products, equipment and chemicals used in an experiment. Model and catalogue numbers may be needed. Consult the journal also for specific instructions if you have sensory data or computer software. It is not sufficient merely to indicate the computer software used to perform numerical computations, although such information may be useful. Specify the mathematical model(s) and statistical methods used to analyse and summarise the data. When appropriate, present an analysis of variance table.

Results and Discussion

Results and Discussion sections may be separate or they may be combined. If separate, then the Results section should contain only results and a summary of *your* research, not a comparison of the literature. It is best to prepare your results first in the form of tables or figures. From among the tables or figures that you would like to report, select those that are the most important or the most representative.

In the results section you explain or elaborate on your findings by logically summarising and illustrating relevant data, using tables and figures (see section "Visuals"). Do not assume it is necessary to report every table or figure that you prepared. Sufficient data should be presented to allow the reader to interpret the results. In the text, describe only the most important results that appear in a table or figure; do not merely repeat numbers from the tables in the text, but rather integrate quantitative data into the text, e.g. "The mean (23.5 g) was greater…". Results that are not significant should be reported in tables or figures, not in the text; if the text becomes too repetitive with similar results, just say that the results were similar. Means and standard errors are an efficient way to summarise in the text a large number of data in a table, but don't report results that are not in a table.

The Discussion section should interpret the results clearly, concisely and logically. For each objective, in order, describe how your results relate to meeting that objective. Cite evidence from the literature that supports or contradicts your results; explain the contradictions. Identify the significant results, and recognise the importance of "negative" or non-significant results. Describe the limitations of your research, e.g. limitations with regard to the design of the experiment, the number of observations, or the statistical analysis. Results, or references to tables or figures already described in the Results section, should not be repeated in the Discussion.

Combining Results and Discussion into a joint section can be a good way to avoid repetition. Make sure, though, to indicate clearly whether you are reporting your own results or results from the literature.

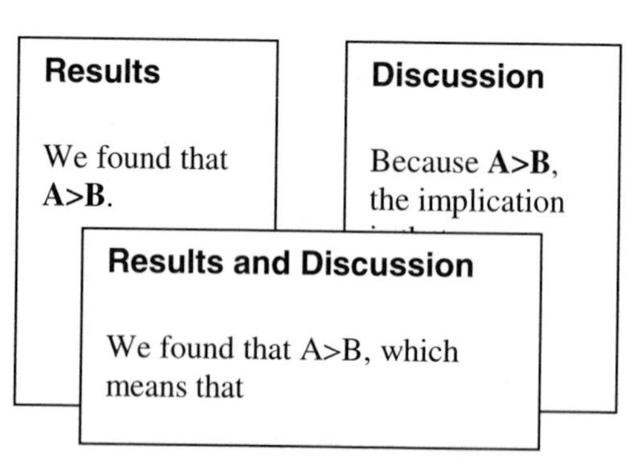

Combining Results and Discussion may avoid repetition

In the Results and Discussion sections, it is recommended that you use the past tense to refer to results of your own research, but you may use the present tense to refer to results that are generally accepted.

Conclusions or Implications

Conclusions summarise the main results of the research and describe what they mean for the general field. This is not the place to mention new results for the first time. Avoid abbreviations, acronyms or citations. Although some guarded speculation is allowed, you should avoid over-extrapolating the results. Implications explain in non-jargon language, or lay terms, what the findings of the research imply. Provide readers with an interpretation of the impact of the research results, where appropriate. Consider suggesting future research to follow up where your research ended.

Acknowledgements

Acknowledgements are often of two types: general and individual. General acknowledgements include those of an institution, of a laboratory, or of a source of funds, whereas individual acknowledgements include those of colleagues and technicians or of an anonymous (or even named) reviewer. If the research is part of your thesis or dissertation, then you might mention it here. If the paper contains a dedication, then this is where it might go. Some journals place acknowledgements as footnotes in the paper, so consult your journal's guidelines.

References

Citations of published literature should follow the instructions set by the journal (or institution) where it is to be published. Instructions vary considerably, as discussed in the chapter "Literature Searching and Referencing". Be certain that all references listed are cited in the text of the paper, and that all references cited in the text are listed.

Appendix

The Appendix provides readers with supplementary material that may not be essential to the understanding of the paper but may be helpful. Such material may include numerical examples (which may instead be included in the text, depending on the nature of the paper), questionnaires, extensive details of analytical procedures, novel computer programs, or derivations of complex mathematical formulas or proofs. Alternatively, supplementary material may be provided on request from an author or on a web site; be sure the address of the author or the web site is available. Consult the journal for the correct placement of the Appendix section, which may vary by journal.

The main sections of a scientific paper can be summarised as follows:

Main sections of a scientific paper[1]

Section	***Intends to tell the reader***
Title	What the paper is about
Abstract	Short summary, which can also "stand alone"
Introduction	The problem, and what is known already
Materials and Methods	What you did
Results	What you found
Discussion	How you interpret the results
Conclusions	Possible implications
Acknowledgements	Who contributed to the work and how
References	How to find the papers referred to
Appendix	Supplementary material

USING VISUALS IN THE WRITTEN TEXT

Tables and figures, such as charts or graphs, help make numbers meaningful and convey information to the reader. Visuals can be used to find relationships, to emphasise material, and to present material more compactly and with less repetition. The number of tables, charts or graphs that you use depends on the amount of information you have and on your purpose in presenting that information.

Tables should be used when you want the reader to focus on specific numbers, e.g. actual data or estimates of parameters; whereas charts or graphs should be used when you want the reader to focus on relationships among those numbers.

---

[1] Not all papers contain every section specified here. In a review paper, for example, the sections Materials and Methods and Results are usually replaced by a thorough literature review, divided into suitable headings.

Consult the instructions of the journal on how to construct a table or a figure (e.g. whether to use leading zeros) or refer to a recent issue of the journal for examples of format. All tables and figures should be referred to in the written text.

Tables

Use tables to summarise numerical values so the numbers can be interpreted logically in the text of the paper (e.g. in Results). The text may refer to a table, but a table should not refer to the text (e.g. an equation in the paper), or to a figure or another table. A table, therefore, should be self-contained, so as to stand alone. The title of the table, for example, should be descriptive enough to stand alone. Use footnotes freely to make the table clear and concise. Make the body of the table concise by avoiding repetitive information, by excluding data that can be computed from available information. Use abbreviations in tables, but if they are not "self-explanatory" add a footnote. Use a logical format by arranging comparisons in columns (vertically) rather than in rows (horizontally) to facilitate mental subtraction or division (align columns by decimal) and by emphasising similarities (group similar items) and differences (separate dissimilar items).

When possible, a table should be organised to fit across the page, so that the page can be read without rotating it; orient the table as "portrait" rather than as "landscape". Avoid the use of vertical lines between columns and use few horizontal lines, except as needed (compare Tables 2:1, 2:2a, and 2:2b).

Make the table legible. Construct a title that explains what the table shows. If in the text you refer to the table, use the same words as in the title of the table, then the reader is certain to be looking at the correct table. To help the reader, round data to simplify. You should never use more significant digits than your method justifies; for example, a mean of 503.753 kg implies that you can weigh to the nearest gram, which is not appropriate if your scales are only accurate to within 1 kg or 10 kg; in this example, the number should be either 504 or 500 kg. Use common units and double-space every five lines for easy reading. Avoid using a dash or zero to indicate absence of results (unless your journal tells you to): zero is a number and could be a valid observation, a dash could be confused with a negative sign. Use "ND" to indicate "no data" or "not determined". Leading zeros (e.g. 0.113) make numbers easier to read and easier to justify in a column, but some journals omit them (e.g. .113).

Table 2:1. An example of a poor table layout

Variable	Breed A		
	Treatment 1	Treatment 2	Treatment 3
1 (units)			
2 (units)			
3 (units)			
	Breed B		
	Treatment 1	Treatment 2	Treatment 3
1 (units)			
2 (units)			
3 (units)			

Table 2:2a. An example of a better table layout

Treatment	Variable 1 (units)	Variable 2 (units)	Variable 3 (units)
	------------	Breed A	------------
1			
2			
3			
	------------	Breed B	------------
1			
2			
3			

Table 2:2b. Another example of a better table layout (when there are more variables than treatments)

Variable	Breed A			Breed B		
	Treatment 1	Treatment 2	Treatment 3	Treatment 1	Treatment 2	Treatment 3
1 (units)						
2 (units)						
3 (units)						
4 (units)						
5 (units)						

Tables are inserted in the text when you write a report or some other kind of paper that is to be reproduced directly from your manuscript. A table should appear near the comment that refers to it. If a table spans a page break, then it is placed on the following page. When you submit a manuscript to be published in a journal, however, tables are usually put in a separate section at the end. This is so the typesetter can format tables to suit the layout of the journal. You should indicate where you want them in the printed paper by inserting the phrase "Table X near here".

Figures

Use only as many figures as necessary to explain the results. Figures should focus on relationships among numbers. Be aware that figures are often reduced in size for publication, which makes it difficult to distinguish between lines. Therefore, use an appropriate scale of lettering, symbol size and boldness of line; scale, size and boldness should be consistent across figures. Do not repeat material already included in another section of the paper (e.g. a table in Results or Discussion).

Figures should be understood independently (stand alone), without reference to the text, to tables, or to other figures. Figures with multiple parts, however, can be designated as 1a, 1b, etc. Abbreviations should conform to the style of the journal and be consistent with the text, and format and style should be consistent with other figures. Each figure must be accompanied by a legend (title and caption) that explains what the figure illustrates. The legend is usually placed under the figure; it may vary, so check the rules for the journal. When you deliver your manuscript to a journal: remember to accompany each figure with a caption, which gives the title and the legend for that figure. Be sure to identify each figure on the back with the figure number, the author's name, and the title of the paper; indicate the top of the figure.

Examples of figures include pie charts, bar charts, line graphs and scatter diagrams (Figure 2:1). Pie charts help to compare a part, or segment, with the whole. Start at 12:00 with the largest or most important segment and go clockwise in some logical order. Limit the number of segments to about five or seven. Label the segments outside the circle.

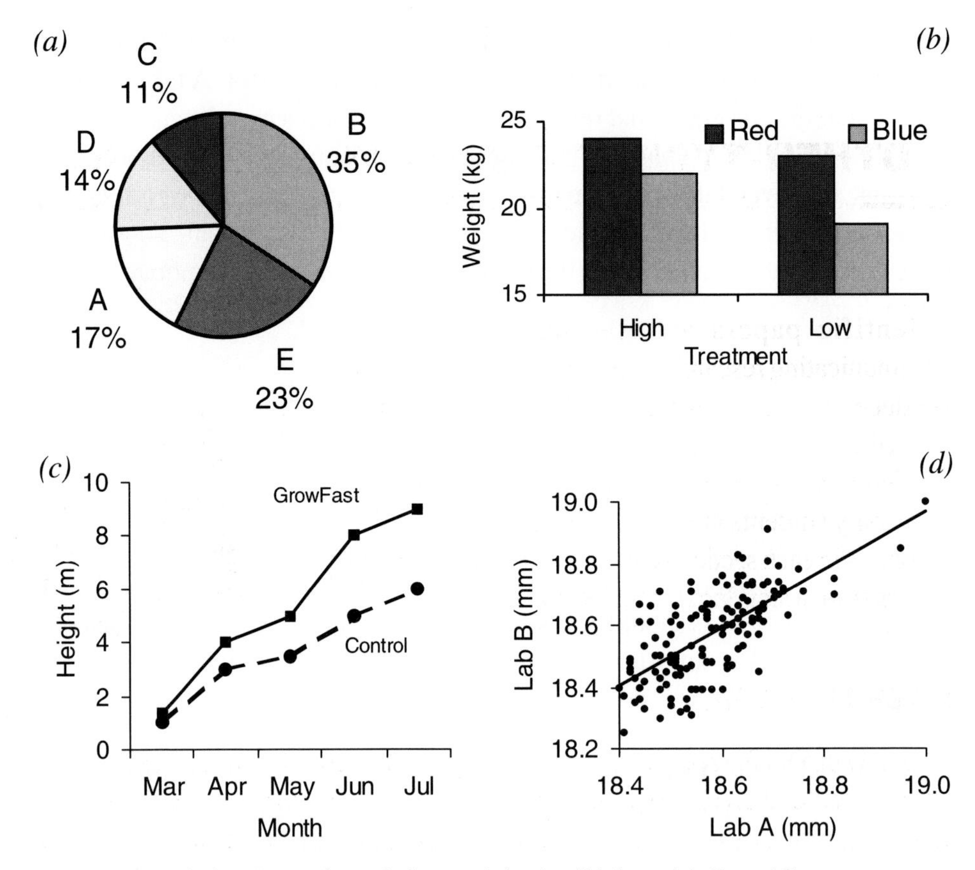

Figure 2:1. Examples of charts: (a) pie, (b) bar, (c) line, (d) scatter.

When comparing one item with another, use a bar chart or a line graph. Bar charts help to compare one item with another or to compare items over time. Order bars in a logical (e.g. chronological) way. Place all labels inside or outside the bars, not both. Make all bars the same width. Three-dimensional figures, especially bar charts, are sometimes difficult to interpret and often obscure the presentation of data.

Line graphs help to compare responses over a continuous scale, such as time or levels of treatment, or to show frequency or distribution. Place time or levels on the horizontal axis. Avoid using more than three lines on a graph, if possible. Use symbols that are easily distinguished, such as triangles, circles and squares; use solid or dashed lines, as dotted lines can appear solid when printed. Scatter diagrams help to show correlation between two variables and, together with the line graph, help to compare actual data with predicted data (regression lines).

3

OTHER TYPES OF SCIENTIFIC WRITING

Scientific papers are the main method for communicating research results, but scientists have to produce many other types of writing. The examples described in this chapter are: literature reviews, conference summaries and abstracts, theses for university students and popular science articles. The emphasis in each section is on how each type of writing differs from a standard scientific paper.

LITERATURE REVIEW

A review differs from a standard scientific paper by reporting work from several sources, rather than one experiment or research programme. Review papers are found in journals and conference proceedings; they are also a common form of writing in university training. Reviews of literature are also found, in a shortened form, in the Introduction section of a standard scientific paper and, in a longer form, in the Literature Review section of a thesis or dissertation. The main requirement of a review is that it should be *critical*. This does not mean that you have to criticise other authors, but that you should compare and contrast information published from different sources. A review is just as much a part of the research process as conducting experiments; good reviews contribute to scientific knowledge by bringing data together so that new, or more definite, conclusions can be drawn. The conclusions from a review of the literature may suggest new areas for research by identifying gaps in the knowledge.

The structure of a review paper is different

As described in the chapter "Sections of a Scientific Paper", a review consists of an Introduction, various sub-sections, Conclusions and a Reference List; there may also be a General Discussion before the conclusions. The Introduction is similar to that of a standard paper; you state what the problem is and why you are going to review the literature, but you don't include the review in this section. The Conclusions and Reference List follow exactly the same principles as those in a standard paper. The major difference is in the body of the review.

The first task is to split the body of your review clearly into themes or topics, each of which can have its own section. Splitting your review into sections should help the reader to follow your reasoning. Keep each topic separate, but put them in a logical order. Starting with general topics and moving to specific topics usually works best, but you might relate specifics back to the general in the Discussion. For example, in a review of burger consumption and human health, you might have the following headings:

1 Introduction (what you are reviewing and why)
2 Influence of nutrition on human health
3 Nutrient requirements of humans
4 Trends in burger consumption
5 Nutrient content of burgers
6 Discussion - burgers in a balanced diet

In this example, we move from the general situation (human health) to a very specific situation (analysis of burgers) and tie everything together in the discussion.

The content should reflect the contrasting results

In a scientific paper, you normally have a hypothesis, which you accept or reject on the basis of your experimental results. In a review, you have a theory or a message, which you support or reject on the basis of published results. You will probably have formed a theory when reviewing the literature (see chapter "Literature Searching and Referencing"), or you may be invited to review a subject for a conference or journal, or you might be writing a review for a thesis or student assignment.

Even if you know the "answer" before you start to review the literature, you still need to convince the reader with sound arguments, supported by good evidence. This applies to all sub-sections and statements in the review; do not simply make a statement that agrees with your idea and give only one reference. Try to have at least two references that support the idea. Even better, give a reference that reports contradictory results and say why it does not fit your theory (e.g. different experimental conditions). You should never omit a relevant reference just because it conflicts with your ideas; equally, you should be cautious when using references that agree with you but are not relevant (e.g. extrapolation beyond the data range).

Never extract parts of a paper that disagree with the original author's conclusions, unless you are looking at the data from a new angle. For example, if Smith (1995) wrote "We found that 99% of birds flew North, the rest flew South", you cannot write "Smith (1995) found that birds fly South". However, if some new studies found that brown birds fly North and white birds fly South, you might say "Colour appears to affect the direction of birds' flight (Brown *et al.*, 2000), which may explain the observation of Smith (1995) that some birds flew South". Do not report results that the author found were not significant as though they were.

Always try to give some experimental details from the paper that shows how strongly the evidence supports your theory. For example, "Jones (1999) found ... in mice; Robinson (2000) found similar results in rats, but only for females".

Quantitative data are useful for supporting theories and formulating new ideas. Rather than writing "Ingredient X increased fuel efficiency", write something like "Fuel efficiency increased ($P<0.05$) from 10.5 km/l in control tests to 12.3 km/l when ingredient X was used". You can combine data from several studies in a table or figure (Table 3:1), but be wary of trying to do statistical analysis on means from different studies; consult a statistician first.

Do not produce a long list of papers to support a minor point; choose some key references that strongly support the point. Try to avoid referring to textbooks; they usually do not describe experimental results, they just state the author's opinions. When you get ideas or support from a paper that is itself a review, always check the original source of information; the reviewer might have misinterpreted the data.

Table 3:1. Example of table layout in a literature review (looking at the height of young people)

Sex	*Number of observations*	*Age range (years)*	*Mean height (m)*	*Source*
Male	250	10-17	1.84	Jones *et al.* (1990)
Female	250		1.76	
Female	100	11-13	1.68	Smith (1993)
Both	312	10-16	1.78	USDA (1995)
Male	96	12-18	1.85	Malik and Olsen (1998)

Writing your review

Like any other scientific writing, writing a review requires planning and careful thought. Read the chapter on "Literature Searching and Referencing". If you follow the advice given there, you will have lots of index cards (or their electronic equivalent) with key words on them. Sort your cards into the order of your review sub-headings. Within each topic, decide how you want to cover the material and sort the cards into the same order. Check whether you have any gaps (ideas without a reference) and fill them by doing a further literature search. Then gather the papers (or photocopies) together in the correct order.

Write the review one section at a time. You may find that ideas can be moved from one section to another; choose which is best, but do not duplicate the same information in two sections. Check each section and make sure that you have evidence to support your arguments (both agreeing and disagreeing, if appropriate); check also that you have quoted each author accurately. When all sections are finished, read through the review and make sure that the flow between sections is satisfactory; you may need to add a sentence that links sections. For further hints, read the chapter "Getting Started in Writing".

Finally, write the General Discussion (if needed), the Introduction and the Conclusions. Make sure that your Introduction and Conclusions use similar words and that the Conclusions refer back to the Introduction. In the Introduction, you might write "xxx is a potential problem"; in the Conclusions, you might write "xxx is only a problem in certain areas of the country …".

CONFERENCE SUMMARY AND ABSTRACT

A conference paper may be in the form of a full scientific paper or a review, but you are usually asked to write a summary and/or abstract. The difference between a summary and an abstract is that summaries are longer, so they can contain more detail, including tables and figures. Abstracts normally stand alone (you don't need to read any of the full paper) and are often published as the conference proceedings. Abstracts might be submitted in advance so that the conference organisers can decide whether to accept your paper and which session to put it in. The purpose of a summary is to support your oral or poster presentation, so that you can concentrate on getting the main messages across to the audience.

You will nearly always be given rules about the length and layout of your summary or abstract. These should be followed carefully, particularly for an abstract, or your paper might be rejected. A conference abstract follows the same rules as it would in a scientific paper – concise, stand-alone, no references or visuals. A conference summary, on the other hand, is usually written with the same section headings as a scientific paper (see chapter on "Sections of a Scientific Paper"), but you do not need to include so much detail in the Introduction and Discussion sections. You do not need a comprehensive review of literature (unless you are only presenting a review of literature) and should just give a few key references if they are directly related to your work. The Materials and Methods section should be similar to that used in a scientific paper, possibly omitting sources of equipment etc. The most important section of your summary is the Results section, which should contain full details of the results that you are going to present.

THESIS OR DISSERTATION

A thesis or dissertation can be produced for a PhD, MSc or BSc degree. The MSc or BSc thesis is usually a monograph, i.e. a complete "book" about your research programme. The PhD thesis is often a monograph, but may, in some universities, consist of a collection of draft or published papers, with an introduction and a discussion to link them together. The main differences between the two approaches are style and layout, but some monograph theses can look very similar to a thesis based on a collection of papers. A thesis is often the hardest piece of writing you will do in your scientific career. Try to spread the load by reviewing the literature and writing up experiments as you proceed through your research programme. See also the chapter "Getting Started in Writing".

A thesis can be a monograph or papers

The Introduction to a thesis should set the scene and outline the approach adopted in the programme. Sometimes an extended introduction is required, which includes a review of literature; in other cases the Literature Review forms a separate section. Check the university regulations and follow the guidelines given in other sections of this book. The section in this chapter on "Reviewing Literature" should help with your literature review. In an ideal thesis you should conclude from the literature that there is a gap in the knowledge; the reader can then turn the page and see that your research has filled that gap.

The main body of a monograph type of thesis is normally split into chapters that correspond to individual experiments (like a collection of papers) or to different aspects of the programme (more like a very long scientific paper). Where each chapter describes one experiment, materials and methods, results and discussion may be included in each chapter, or general chapters may be written for methods and discussion that are common to several experiments. The candidate should choose which layout to use, in consultation with the supervisor, but often the layout is determined by the nature of the programme. There is no right or wrong way of dividing a thesis into sections.

The General Discussion for a thesis or collection of papers requires a slightly different approach from that in a scientific paper or review of literature. You still need to compare your results with previously published information, and discuss the implications, but you also need to discuss the relationships among your individual experiments and state how the overall programme fits your hypothesis. In other words, you must consider the whole picture as well as the individual pieces.

POPULAR SCIENCE

Research results need to be communicated, not only to scientists, but also to public audiences, including all those people involved in the implementation of results. If this communication is not done effectively, the new findings might not be spread and contribute to developments, and the public might feel anxiety and fear about what is going on in research. Scientists, therefore, have great responsibilities in communicating their discoveries, but often need training in how to do it effectively.

Research results should not remain embedded in the scientist's world

Writing a popular science article is different from writing a scientific paper, a review paper or a conference paper. Some characteristics of a popular science article were outlined in the chapter "Communicating Science" (Figure 1:2). We will now discuss in more detail what it might be important to consider when writing a popular science article.

Adapt to your audience

Adapting to the readers' previous knowledge and experience is of utmost importance when writing a popular science article. Your writing will be different if the target group is people directly applying your research findings (such as technicians, farmers, etc.), than if you address an audience with no previous experience in the area. The readers must be able to relate what you say to their own world, or they will not catch your message or not even read the article. So, analyse your aim and your audience; think of the questions who-why-what-how when planning your popular science article (see chapter "Communicating Science").

The subject matter affects your potential to attract readers, especially if you address a public audience. Your topic needs to be interesting *per se*, and may arouse more interest if it is a current topic where clarification is needed.

Simplify results – draw conclusions

Your popular science article should be focused on its topic. Therefore, give the general overview! Avoid presenting too many details, and make sure the information is put in context. Follow the communication ABC, i.e. be accurate, brief and clear, but give sufficient explanations for the reader to understand. Remember that the purpose is not only to communicate facts, but also to make them comprehensible and interesting, to discuss their possible application and impact, and to make the readers feel involved.

I didn't ask for all the details

One of the tasks in writing popular science is to simplify results so that you report only what is important for your message and omit the fine detail. This does not mean that you report only positive results; it is vital that you still give a balanced and honest report of your findings. In a scientific paper you would give full details of statistical analyses, such as regression coefficients and standard errors, whereas in a popular science article you might report only the means, or you might just describe the differences between treatments with very few numbers. The process of simplifying results is similar to writing an abstract, except the words are rewritten for a popular audience.

Apart from simplifying the results, you should also omit most of the detail concerning materials and methods; your experimental design and techniques will not interest the lay reader, and might detract from your real message. Make sure to report your conclusions and implications, however, since these normally form the key message in popular science.

If you cannot decide how much information to include, whether your message is understood, or how to write something more simply, use a live audience. Try to explain your topic to a complete novice, such as your mother, your spouse, a friend, or to a colleague from another discipline. You will have to use very basic language to do this, and make your message simple; this will give you a good foundation for building your article.

Organising the popular science article

A popular science article should have a structure that makes the reader want to read it immediately. Therefore, consider especially the:

- Title
- Preamble
- Headings
- Visuals
- Layout

The title is of vital importance for capturing the readers' interest; so compose a title that is both exciting and informative. The title must be short; it will often be a statement, but it might also be a question. The tense used is usually the present tense.

A popular science article often has some introductory text – a preamble – that is separated a little from the body text, in order to be seen better. The preamble should make the reader curious and interested enough to continue reading; it should add some vital information, not just repeat what is said in the title. The preamble is not like the abstract of a scientific paper, which summarises the main parts of the content; rather it should emphasise the importance of the topic, and give some hints about the content of the article to raise curiosity.

The body text of a popular science article does not follow any standard structure; the text may be divided into sections, each having an introductory heading. These headings are important tools for attracting readers; they make the text less compact, and they help readers to quickly see what the section (and the article) is about; they are like a road map! So, create headings that are not only informative, but that are also eye-catching. Remember, however, that the headings should be short.

Visuals can be very effective in attracting readers, conveying a message and making complicated issues more understandable. Use photographs, drawings, clip-art, tables, and graphs, such as simple diagrams, bar charts and pie charts (see chapter "Sections of a Scientific Paper"). Because visuals capture readers' attention, the figure legends can also be used to emphasise important messages. A legend should explain what is shown in the visual, but not repeat what is self-evident. The legend might contain narrative, drawing some main conclusions from what is shown, for example, so that a reader who only looks at the title, headings and illustrations, will still catch the essentials. Explanations can also be given within a visual.

Layout might be the main factor that encourages the reader to take a closer look at your article. The style of the magazine usually determines the final layout and length, but you can usually make suggestions. Remember that your text will more likely be read if it is not too long. Before writing your popular science article, look at the style of articles previously published in the same magazine.

Use language that is easily understood

The readers of a popular science article are usually not familiar with the language used in your scientific area. Therefore, use words that are easy to understand, and explain scientific or technical terms if you have to use them. You will find many ideas on how to make your language easier to read and understand in the chapter "Improving your Writing".

Not everybody understands Greek

The tone of your writing can be more personal in a popular science article than in a scientific paper. What you say should stimulate the reader, and your involvement with the topic is important for that to happen. Use relevant examples and analogies to help the reader understand your topic. Furthermore, make logical transitions between sentences, paragraphs and sections, and check your text for "flow and thread". See also the next chapter "Getting Started in Writing".

4

GETTING STARTED IN WRITING

Many of us feel resistance when the time comes to start writing. The task may seem overwhelming, our thoughts circle around and around. We try to grasp the document as if it were to be written in one single step. It is hard to see the structure and it is difficult to find the words. But don't despair, a research paper or report is normally not completed in a single step. Your writing can be divided into several stages, and thus be made easier to handle:

- Analyse your aims and your audience
- Make tables and graphs of interesting results and decide what messages to communicate
- Make an outline
- Write a draft – start with the easiest parts
- Revise and edit

Whether you write a popular or a scientific paper, or a report, such as a thesis, think of the questions who-why-what-how. What is your purpose in writing? What results do you want to communicate, and why are they important? What are the most likely questions that might be raised by your readers? Having the readers in mind throughout your writing will not only make the text more interesting; but it will also give you motivation and make the writing more vivid.

When you carry out research, start writing early in the research process. Don't wait until all results are in place, or all analyses are performed. Put results together in tables and graphs as you go along. Then you will get inspired, and you will know better what there is to write about. Starting writing early might even help you to identify missing parts in your research.

Chapter 4

MAKING AN OUTLINE FACILITATES WRITING

Carefully reflect upon what to include in your document, and make a preliminary outline. This will help you to organise your thoughts, to structure the text so that it is logical and clear, and to remember important parts. The outline facilitates splitting the writing into steps and sub-steps, without losing the overview and the thread through it all.

Choose your own way to prepare an outline

An outline can be prepared in different ways. It might consist of a structured order of headings and subheadings, with key words under each, but it might also be written as an organisational chart or a mind map. The most important thing is that it should be a working outline, a tool to help your writing. It is a plan for you, not for the reader. Your outline will not be complete from the start. It can, and normally should, be revised continuously during your writing. Once you have an initial outline for the document, you can easily add key words or references to notes or literature.

Before starting to write your manuscript, discuss your proposed outline with other people, such as your co-authors, a colleague or, if you are a student, with your supervisor. This will help you to find the best approach and it will save time, as it will hopefully reduce the need for a large revision of the structure and content of your document at a later stage.

Structure the text

The answer to the question, how to structure a text, is not simple. The structure differs depending on the type of text you are writing. Examples of structuring principles are:

- Chronological order (development over time)
- Order of interest/importance (the most important first)
- Cause and effect (or the opposite order)
- Comparison/contrast ("pros and cons")

When writing a scientific paper or report, you are normally compelled to include sections such as: Abstract, Introduction, Materials and Methods, Results, Discussion and Conclusions, but within those sections you can choose the structure. If you write a popular science article you are usually not tied to a specific format, but you need to structure the article so that the readers get interested immediately. Whatever kind of structure you use for your writing, make sure that it is logical and clear. Using different types of structures for different parts of a manuscript may sometimes be the best solution.

USE A COMPUTER FOR YOUR WRITING

Composing text directly at the computer is an efficient way to facilitate the writing process. Some of the main advantages are:

- It's easy to revise text and lay-out
- Knowing that it's easy to revise may have a positive "psychological" effect on your writing
- Text, tables, figures etc. from various sources can be merged
- Continuous over-view of text on screen and in paper copy

Using word-processing doesn't change the need for thinking and planning in writing. Because your manuscript doesn't need to be "perfect" from the beginning, however, it is easier to get started in writing. You can also use the computer to produce the visuals you want to include in your document. Sometimes you can do this within your word-processing programme, or you might need specific software for making graphs and drawings.

Valuable options in word-processing programmes

If you are not already familiar with the major options in your word-processing programme, it's worthwhile learning them. It might also be useful to learn some short-cut commands, so that you don't need to mouse-click too often.

In word-processing, you can add or delete text easily. By using the *cut* and *paste* options you can move words, sentences, paragraphs or entire sections, and you will see the result on the screen immediately. If you don't like the result of a change you can easily *undo* it. You can *replace* one word with another throughout the manuscript. Construction of tables can be facilitated by use of the *table* option, and tables and figures can be placed just where you want them to be in your document. Note, however, that tables and figures are usually enclosed separately when a manuscript is delivered to a publisher. Numbers and text can be *sorted* in desired order. The *format* can be changed easily with regard to font type and size, bullets, indents, line spacing, margins, number of columns, page numbers. A *contents list* can be produced automatically from your headings, and you can also produce an *index*. Advanced formulas and equations can be written with ease, although you may need additional software.

Spell-checking can be done for a large number of languages, either automatically or on your command. Be aware, however, that you still need to proof-read your document carefully, especially for coherence. The spell-check doesn't function if a typing error from your side has resulted in a word that exists, although there is an error in context, such as an error in the use of to and too. *Grammar check* can also be helpful, but it sometimes disagrees with constructions that are required in scientific language, e.g. passive voice. Another useful tool is the *thesaurus*, which helps you find synonyms to vary your language. It you need to produce an abstract that stays within a specified number of words or characters, you may find the *word count* option helpful. You can also *track changes* that are done in your manuscript by a co-author, for example.

The list of options in word-processing can be made much longer, and new options will be added all the time with new versions of software. The best way to learn the options is to test and practice, maybe read a book on the topic, or discuss with people who are experienced in using word-processing. In any case, remember to ***save*** your file frequently during the writing session, and to ***back-up*** your document on additional discs. It is also wise to make a printout.

START WITH THE SECTIONS YOU FIND EASIEST TO WRITE

Having produced some tables and graphs, and also an outline, it is time to start writing. You know, or have an initial idea of, which sections and subsections you plan to include. Therefore, you don't need to write your document straight through from beginning to end. It might be a better idea to start with the sections that you find easiest to write. This is valid for most situations of scientific writing, although we will focus here on a scientific paper based on your research.

Writing a paper is like building a house

Writing a scientific paper can be compared to putting pieces together when building a house. Before you build the walls and rooms (materials, methods, results, tables, figures), you need a plan (outline) and a foundation (basic knowledge). The door and the windows (introduction and outlooks) are produced separately and put in place later. The house is kept together by the roof (discussion), ending up with a chimney and smoke (conclusions) spread widely. For a good structure, the separate parts must be adjusted to fit together, just like logical transitions are needed throughout your paper. Once the house is in place, it is simple to make an overview picture of it (abstract).

Because you can build your scientific paper in units, it is easy to start planning or writing parts before the research is completed. Doing so means that you think of the paper early in the process. This may give ideas and thoughts that can continuously be put on paper, at least as key words or short notes. Writing takes time, so the sooner you start the process, the better!

You choose where to start

You might start to write a single section or you might choose to write several sections in parallel. You can always write a short first draft for an Introduction, explaining why your subject is important, what is known about it and what is not known, as well as what are the objectives of your study. This might be helpful when you write the results, as they should comply with the objectives. The Materials and Methods section is often considered easy to write, and could also be a good starting point.

For the Results section: use the tables and graphs that you have put together previously, and extract the most interesting parts to enter into tables and graphs for your paper. Build from that when you start writing about the results, but before then it might be useful to explain them to someone else. By doing so, you will begin to get an overview of your results and also a basis for selecting which tables and figures to include in the paper. It might also help you to prepare for writing the Discussion part and in formulating your conclusions. When writing a specific section, however, think of it as part of the full paper and its role in the entirety. Make sure, also, that what you write will be in line with the objectives given in the Introduction.

WRITE A DRAFT – REVIEW AND REVISE

Before writing the first draft, it is a good idea to check which editorial rules and format you need to follow, such as how to prepare tables and figures, or how to refer to literature in the text and in the reference list. Doing these things correctly from the beginning will save you time and frustration.

Get words down

In the first draft of your text, strive to get the words down. Don't feel pressured to make a "perfect" text; merely think of the content that you want to describe and discuss, and write it! It is just a first draft, and you will revise and edit the text in future drafts. Imagine your readers asking you questions, such as: why is this important, what did you do, what did you find out, how can you explain that, what use can we make of it, and why are you telling me that? Remember to be accurate, brief, clear and logical in your writing. This also means that you should emphasise what is most important. What comes at the beginning of a sentence, a paragraph, a section, a chapter or the full text is usually considered to be the most important. Furthermore, it might be worthwhile to have a look once again at the advice given in the chapters "Sections of a Scientific Paper" and "Improving Your Scientific Writing" before you start writing!

Prepare for the next writing session

Before closing a writing session, write some key words on what you will write about in the coming part, or even write the first sentences. Doing so will help you overcome the initial resistance you might feel to getting started next time. If you find it difficult to express some of your thoughts in writing, explain them to someone with the same background as your audience to get feed-back on what is understood and what is not. Then you'll find it easier to express the matter next time you sit down to write.

Express your thoughts orally if you get stuck when writing them down

Revise and edit – let somebody read your draft

Revision of the text is usually done in several stages. The first revision might best be completed directly at the computer. Read through each paragraph, asking yourself: "Am I saying what I mean?" and "Would I understand this if it were new to me?" Make sure that pronouns (such as "it" and "this") refer to the right noun. Revise as needed to make the text clearer. Then make a first quick check of the text for fluency of reading, coherence, and for typing errors. Having done this, it might be a good idea to leave the text for one or two days, so that you look at it with "fresh eyes" at the next revision.

In the second revision (probably best done from a printout) you might focus on the content and structure of your document. For example: Is the title relevant? Are all compulsory headings included, and in the correct order? Will the abstract "stand alone", i.e. can it be understood without reading any other parts of the paper? Is any important content missing? Does the text agree with the tables and figures? Can the tables and figures be further improved to be clearly and easily understood? Is the source given where appropriate? Is the reference list in accordance with the references cited in the text?

Having done this much revision it will be wise to ask one or a few people to read your draft (compulsory, of course, if you have a co-author(s) or a supervisor). This can be quite valuable and will help to get a view from other angles. It is recommended that you ask these people to read your draft once again, at a later revision of your paper. The more complete your text is, the easier it is to judge how well it can be understood and if something is lacking.

At some later revision, we suggest that you read aloud your full text straight through from a paper copy. Make sure that the various parts fit together well;

that you see the "thread" through it all. Check that the transitions between sentences, paragraphs and sections are logical. Delete unnecessary repetition, unless it has been done intentionally to emphasise an important message. Furthermore, is the tense chosen appropriate? Without losing the flow of your reading, indicate in the margin or directly in the text where revision is needed. Go back to the computer and do the revision.

When you feel that your text has "flow and thread", do yet another revision. Check for redundant words, long sentences, possible synonyms, etc. Make sure again that your manuscript follows the editorial rules and format of your publisher. Renew the spell-check after each revision, or ever better, let the function be on continuously. Do a final check that references in text and the reference list are in agreement; you might have added or deleted references during your revision.

If you write in a foreign language, a linguistic revision by a native-speaking person might be the last step in preparing your manuscript.

5

IMPROVING YOUR WRITING

Vigorous writing is concise. A sentence should contain no unnecessary words, a paragraph no unnecessary sentences, for the same reason that a drawing should have no unnecessary lines and a machine no unnecessary parts. This requires not that the writer make all sentences short, or avoid all detail and treat subjects only in outline, but that every word tell.

The Elements of Style (2000)
William Strunk, Jr. and E. B. White

The items we will discuss in this chapter include choice of words, tight writing, sentence length and structure, transitions to link ideas, words that are often confused, using correct grammar and punctuation. Many of the examples are taken from Locker (1999) and further help can be obtained from the books in the reading list.

We hope that this chapter will make it easier and more enjoyable for you to write an effective scientific paper. But first, a couple of hints about writing your paper: before you start writing the paper, prepare a one- or two-page outline, including the contents for each section, and, after you have finished writing the paper, be sure to proof read it and check the spelling! For further suggestions, see the chapter "Getting Started in Writing".

TEN WAYS TO MAKE YOUR WRITING EASIER TO READ

Writing a scientific paper can be easy and even fun. One way to make it easy is to decide on a set of rules that will guide your writing style (e.g. use *If …, then …* sentences when appropriate), and then adhere to those rules. For example, if you follow rules as your write, then many of the decisions about style are ready made. Your rules may change over the years, as you change your writing style; that is OK, and even desirable. If writing is easy for you, then your writing will be easy to read. Examples of ways to make your writing easier to read are as follows:

1. Choice of words:

Use words that are accurate – that mean what you want to say; words that are appropriate – that fit well with other words in the paper; and words that are familiar – that are easy to read and understand. Use specific, concrete words; they are easier to understand and to remember.

When you have something simple to say, say it simply. Use the word that conveys your meaning most accurately, but when deciding between two such words, choose the shorter, more common word. There are exceptions to that rule, however. Use a long word if it is the only word that expresses your meaning accurately, if it is more familiar than a short word, if its connotations are more appropriate, or if scientists in your discipline prefer it.

For example:

Instead of ...	*Use ...*
ameliorate	improve
approximately	about
commence	begin
enumerate	list
finalise	finish, complete
prioritise	rank
utilise	use
viable option	choice

2. Avoid jargon:

Use technical words and expressions (jargon) only when the terms are essential and familiar to the reader. Otherwise, avoid the use of jargon because it is difficult to understand. Instead, use a simpler "plain-language" equivalent, even when the equivalent expression is longer.

3. Use active verbs and avoid passive verbs:

A verb is **active** when the subject does the action. A verb is **passive** when the subject is acted upon. To identify a passive verb in a sentence, look for a form of the verb *to be* followed with *by*. For example, "This method *was* recommended *by* them" is passive, whereas "They recommended this method" is active.

A sentence that uses an active verb is shorter and clearer; more interesting and less boring; more direct because it emphasises the subject; more forceful; takes less time to read and is easier to understand; and sounds less pompous and bureaucratic.

Passive verbs are desirable to provide coherence within a paragraph, i.e. to provide transitions between sentences by repeating a word. For example, "These problems ended with the discovery of penicillin. Penicillin now could be used by ...." A sentence is easier to read if "old" information comes at the beginning of a sentence. Start one sentence with the idea that ended the previous sentence, even if it means using a passive verb.

Your choice of words is important!

It is desirable also to use the passive verb in the following situations: to emphasise the receiver of the action, e.g. "Watson and Crick were awarded the Nobel Prize"; to avoid assigning blame, e.g. "It is known that there are errors associated with field data"; or to omit an unknown or irrelevant agent, e.g. "The petri dish was warmed slightly".

4. Use strong verbs – not nouns:

Put the weight of the sentence in the verb. Strong verbs make sentences more forceful and easier to read. Instead of writing, "We performed an analysis of the data," it is more forceful to write, "We analysed the data." Nouns ending in *-ment*, *-ion* and *-al* often hide the verb.

For example:

Instead of using the noun in ...	*Use the verb ...*
make an adjustment	adjust
perform an analysis	analyse
provide assistance	assist
reach a conclusion	conclude
take into consideration	consider
make a decision	decide
perform an investigation	investigate

5. Tighten your writing:

If the same idea can be expressed in fewer words, then the writing is wordy. Wordy writing bores the reader and makes it difficult for the reader to understand what you mean. Good writing is tight, and tight writing allows you to convey more information. Tight writing is important, especially when you have to write an abstract with strict limits on the numbers of words or characters allowed.

To tighten your writing, follow these strategies:

- eliminate redundant words whose meaning is already clear:

  ~~a period of~~ three months; during ~~the course of~~ the experiment; during ~~the year of~~ 1999; maximum ~~possible~~; ~~past~~ experience; plan ~~in advance~~; refer ~~back~~; ~~the colour~~ blue; ~~true~~ facts; repeat ~~again~~.

- eliminate words that say nothing:

  it is interesting to note that; quite; really; rather; the (especially with plurals); very (if it substitutes for damn!)

- substitute a single word for a wordy phrase:

Instead of using ...	*Use ...*
at the present time	now
due to the fact that	because (NOT since)
it may be that	perhaps
in the event that	if
in the near future	soon
prior to the start of	before
on a regular basis	regularly
a second point is	secondly
more often than not	usually
would seem to suggest	suggests
one of the problems	one problem
in spite of the fact that	although, despite, or nevertheless
on two separate occasions	twice
were found to be in agreement	agree
take into consideration	consider
carry out experiments	experiment
it is obvious that	obviously

- use infinitives (the *to* form of the verb; e.g. to run) and gerunds (the *-ing* form of a verb; e.g. running) to make a sentence smoother and shorter
- combine sentences to eliminate unnecessary words and to focus attention on key points
- put the main idea of your sentence into the subject and verb to reduce the number of words

Think about what you *mean* to say, write it in different ways, and choose the tightest one.

Phrases beginning with *of*, *which*, or *that* often can be shortened. Instead of writing, "The estimates of the parameters were . . .," it is shorter to write, "Parameter estimates were ...." Sentences beginning with *There are* or *It is* can often be tightened. Instead of writing, "There are three reasons for these results", it is tighter (and stronger) to write, "Three reasons for these results are ...".

6. Sentence length and sentence structure:

Keep sentences short and simple. Simple sentences have one main idea. Compound sentences have two main ideas that are closely related, and they are joined with conjunctions such as *and*, *but* or *or*. Complex sentences have one main idea and one subordinate idea related logically, e.g. "If ..., then ...".

Always edit sentences for tightness, but use long sentences to link ideas; to avoid a series of short, choppy sentences; or to reduce repetition. When using long sentences, keep the subject and verb close together by putting modifying material at the end of the sentence. Instead of starting a sentence with a reference to some previous research, for example, start with the main finding of the research and place the reference at the end of the sentence. Strive to put the most important idea early in the sentence.

7. Use parallel structure:

Use the same grammatical form or consistent pattern for ideas that have the same logical function. Parallel structure makes writing smoother, more forceful, and easier to understand, especially when used for a list or a series of ideas. Be sure that each item in the list begins or ends with the proper word form.

Parallel structure is especially useful when writing results and the discussion. Once you decide on the form of the structure to present the result for one variable, say, use the same pattern to present results for other variables; simply copy, paste, and revise the text. Having mastered the result for one variable, the reader can see the pattern and can understand results easily for other variables.

8. Use transitions to link ideas:

Transition words and phrases (e.g. *and*, *for example* or *during*) signal connections between ideas. Transitions tell the reader if the next sentence continues the previous idea or starts a new one. They tell the reader whether the idea that comes next is more or less important than the previous one. Transitions are used also to introduce the last or most important item, to introduce an example, to compare or contrast ideas, to show cause and effect, to show time, or to summarise or end. The following are examples of transition words and phrases:

- To show addition or continuation of the same idea:
  and, in addition, also, likewise, first, second, third, similarly
- To introduce the last or most important item:
  finally, moreover, furthermore
- To introduce an example:
  for example (e.g.), to illustrate, for instance, namely, indeed, specifically
- To contrast:
  in contrast, on the other hand, or, whereas
- To show that the contrast is more important than the previous idea:
  but, nevertheless, however, on the contrary
- To show cause and effect:
  as a result, for this reason, because, therefore, consequently
- To show time:
  after, next, as, then, before, until, during, when, in the future, while, since
- To summarise or end:
  in conclusion, to summarise

9. Write coherently:

Coherence refers to the logical sequence of sentences within a paragraph. Just as you should begin a sentence with the subject, which states the main idea of that sentence, you should begin a paragraph with a topic sentence, which states the main idea of that paragraph and provides structure to the paper. A good topic sentence forecasts for the reader the content of the paragraph and holds the paragraph together. A paragraph that lacks a topic sentence, lacks unity. If a paragraph contains more than one main idea, consider linking ideas with a transition sentence. Otherwise, consider splitting the paragraph into more than one.

To improve coherence within a paragraph, discuss only one idea, or one topic, at a time. Use the same organisational pattern for successive sentences. Use parallel structure for the main subjects and main verbs. Tell the reader what to expect, e.g. "There are two problems with this method: the first is ..., and the second is …"; then go on to elaborate on the problems.

10. Make your logic clear:

Write what you really mean to say and write it logically! If you have difficulty putting an idea onto paper, say "What I Really Mean To Say Is ...," and write the words you mean to say. Once the words are on paper, you can revise them. Above all, make sure what you write is coherent and makes clear, logical and scientific sense!!

HALF-TRUTHS ABOUT STYLES

It is important to organise a scientific paper according to the standard format for the journal. It is also important to write the paper in a style that uses clear and concise language, so as to allow the reader to understand your research paper thoroughly and efficiently.

As you develop and improve your writing skills, you will follow certain rules about style: some are truths and some are half-truths. Truths are general statements about style that are always true (e.g. start a sentence with the main idea, a subject and verb must agree in number), whereas half-truths are general statements that are only partly true, and they should be applied carefully. Half-truths about style are:

- ***Write as you talk***: Read what you have written aloud. If it sounds awkward, revise it. "Writing as you talk" is OK for the first draft (to get something on paper, especially if you are faced with "writer's block"), but then revise it for a better style.

- ***Never use I or we***: Using *I* or *we* when you write about things that you did in the Materials and Methods section may be more appropriate and smoother than using passive voice or using awkward phrases, such as *this writer* or *these authors*. Some journals, however, may object to this style of writing.
- ***Never begin a sentence with* and *or* but**: If you begin a sentence with *and* or *also*, then it appears to the reader that the next idea is an afterthought. Instead, embed *also* at an appropriate place in the sentence or instead use *moreover* or *furthermore*, which can also be embedded in the sentence. If you begin a sentence with *but* or *however*, it tells the reader that the next idea is in contrast to the previous idea. It is better to embed the word in the sentence, however, so as to begin the sentence with the most important idea, the subject.
- ***Never end a sentence with a preposition***: The end of the sentence, like the beginning, is important and should be emphasised. Prepositions, e.g. *with*, *in*, *under*, *at* or *for*, are usually not worth emphasising and, therefore, should not be at the end of the sentence. One expects something to follow the preposition, but nothing does when the preposition is at the end. Choice of preposition is sometimes difficult so, when in doubt, use *for*; it almost always works.
- ***Big words impress people***: The purpose of writing a scientific paper is to inform your readers, not to impress them or to show off that you know big words, which only results in making your writing sound pompous. Big words used incorrectly are embarrassing and make you look foolish. Big words used correctly, however, can convey information effectively. Don't hesitate to use a big word or expression, therefore, if it conveys the appropriate meaning clearly and concisely.

WRITING CORRECTLY

Everyone, including native and non-native speakers of English, makes errors while writing. For effective writing, it is best to avoid making errors. Common errors involve issues of spelling, grammar, punctuation and word choice. Errors of spelling can be corrected easily using spell-checker software, but there are dangers: if the error spells another word, it will not

Use the correct dictionary

be corrected; if you use the wrong dictionary (especially US versus British), you may introduce errors instead of correcting them. Errors of grammar, punctuation and word choice are more difficult to correct (grammar-checking software may give good suggestions, but often cannot cope with scientific writing). Examples of common errors and some suggestions for avoiding them follow.

Grammar

Grammar is one aspect of writing that writers seem to find most troublesome. Six common issues of grammar include:

Agreement

Be sure that subject and verb agree in number, i.e. that they are both singular or both plural. Use a plural verb when two or more singular subjects are joined by *and*. When the sentence begins with *Here* or *There* be sure the verb agrees with the subject that follows the verb (e.g. *Here is* or *There are*). Some words that end in *s* (e.g. *series*) require singular verbs. When proof-reading your paper, find the subject and the verb and make sure they agree. Be especially mindful that the word *data* is plural and requires a plural verb, e.g. *are* or *were*.

None can take a singular or a plural verb, depending on the sense and the sound. In the context of the sentence, if it seems like a singular, use a singular verb; if it seems like a plural, use a plural verb. The word should not cause you to stumble during the sentence when you read your paper aloud.

Be sure also that the noun and pronoun agree in number. The following words require a singular pronoun: *everybody, each, either, everyone, neither* and *nobody*. If a situation does not follow the rule, or if the rule results in an awkward sentence, revise the sentence to avoid the problem.

Each pronoun must refer to a specific word, without ambiguity. If a pronoun does not refer to a specific word, add the word to make the sentence clear and unambiguous. Make sure especially that *this* and *it* refer to a specific noun in the previous sentence (the antecedent). By default, the antecedent is the noun mentioned last. If the antecedent is not the noun being referred to, add the correct noun. Use *who* or *whom* to refer to people; use *which* to refer to objects. *That* may refer to anything: people, animals, organisations or objects.

Case

Case refers to the role that a noun or pronoun plays in a sentence. Use nominative pronouns (e.g. *it, they*) for the subject of a clause. Use possessive pronouns (e.g. *its, their*) to show who or what something belongs to. Use objective pronouns (e.g. *it, them*) as objects of verbs or prepositions. Use reflexive and intensive pronouns (e.g. *itself, themselves*) to refer to or to emphasise a noun or pronoun that has already appeared in the sentence.

Dangling modifier

A modifier is a word or phrase that tells more about the subject, verb or object. A **dangling** modifier is a word or phrase that modifies a word that is not in the sentence. Don't write, "After closing the pen, the animal was fed," when you mean to write, "After closing the pen, the technician fed the animal." Note that the technician is doing the feeding, but the technician is not in the original sentence (so the animal might have closed the pen!). To correct the dangling modifier, recast the sentence to put the subject first, "The technician fed the animal after closing the pen." When using a verb or adjective that ends in *-ing*, like *closing*, be sure it modifies the subject of the sentence, *technician* not *animal.*

Misplaced modifier

A **misplaced** modifier modifies an element of the sentence other than the one intended. Instead of writing, "We only measured two samples," write, "We measured only two samples." To correct the misplaced modifier, therefore, move it closer to the word it modifies or add punctuation to clarify the meaning.

Parallel structure

Items in a series or list that are the same in content and form must have the same grammatical construction; it is **not** easier for the reader if you vary the construction. The same construction, or parallel structure, makes it easier for the reader to recognise the similarity of content and form. Instead of writing, "The tissues were extracted, fixed and staining was performed," write, "Tissues were extracted, fixed and stained." An article or preposition that applies to each item in a series must be used only before the first item or else must be repeated before each item; for example, "Tissues were extracted, were fixed and were stained."

Correlated expressions, such as *both…and…*; *not only…but also…*; *not…but rather…*; *either…or…*; *neither…nor…*; *first…, second…, third…*, must be followed by the same construction. For example, write, "This result means not that …, but rather means that …."

Predicate errors

The predicate of a sentence, the part that refers to the subject, must fit with the subject. In sentences using *is* or other linking verbs, the complement must be a noun, an adjective or a noun clause. Instead of writing, "The reason for this result is *because* ...", write, "The reason for this result is *that* …". Be sure that the verb describes the action done by or done to the subject.

Punctuation

Punctuation helps the reader know what to expect. Punctuation takes the reader through the sentence and from sentence to sentence. Use correct punctuation, as in the following examples, to help you avoid common errors in writing.

Comma splices. A comma splice occurs when two main clauses are joined only by a comma and not by a comma and a conjunction. Correct a comma splice by subordinating one of the clauses or by adding a co-ordinating conjunction, such as *but*. For example,

☹ "The experiment began in June, the exact date was unknown."

☺ "The experiment began in June, but the exact date was unknown."

Correct a comma splice also by using a semicolon, if the ideas are related, or by starting a new sentence, if the ideas are not related. For example,

☺ "The experiment began in June; the exact date was unknown."

Run-on sentences. A run-on sentence joins several main clauses only by a conjunction, such as *and, but, or, so* or *for*, and not by a comma and a conjunction. Correct a run-on sentence by adding a comma or by separating a long run-on sentence into two sentences. For example,

☹ "The experiment began in June but the exact date was unknown."

☺ "The experiment began in June, but the exact date was unknown."

Sentence fragments. A sentence fragment is a group of words that is not a complete sentence, but that is punctuated as if it were. Correct a sentence fragment by adding the missing parts of the sentence or by incorporating the fragment into an adjacent sentence. For example,

☹ "By monitoring the animals, we measured two things. The amount of water they drank. The amount of feed they ate."

☺ "By monitoring the animals, we measured two things: the amount of water they drank and the amount of feed they ate."

☺☺ "We measured intakes of water and feed for each animal"

Apostrophe (') Use an apostrophe in a contraction to indicate that a letter is omitted, e.g. *it's* for *it is*; to indicate possession, e.g. the person's DNA; to make a plural, but only if it could be confused for another word, e.g. A's. With dates, 1990s means the decade 1990–1999; 1990's means belonging to the year 1990.

Colon (:) Use a colon to separate a main clause from a list or to join two independent clauses when the second clause explains or restates the first.

Comma (,) Use a comma to separate the main clause from an introductory clause or words that interrupt the main clause. Do not use a comma, however, to set off **essential** information. Use a comma after the first clause in a compound sentence if the clauses are long or if they have different subjects. Do not insert a comma simply because you stop to take a breath. Use a comma to separate items in a series. Consult the journal instructions to learn if you are allowed to place a comma before the *and* and/or *or*.

Dash (—) Use a dash to emphasise a break in thought.

Ellipsis (...) Use an ellipsis to indicate that one or more words have been omitted, usually within quoted material.

Full stop or period (.) Use a full stop at the end of the sentence or after some abbreviations.

Hyphen (-) Use a hyphen to join two or more words used as a single adjective, e.g. two-day old baby or two day-old babies.

Parentheses or brackets ( ) Use parentheses to set off words, phrases, or sentences used to explain the main idea, e.g. The maximum daily gain (10.5 kg) was achieved by ....

Quotation marks (“ ”) Use quotation marks around words that you think are misleading or around words that you are discussing as words. Use quotation marks around words that you quote directly from someone else. Journals have different rules about single (‘) or double (“) quotation marks, so check their instructions; one convention is to use double marks for quoting someone and single marks for emphasising a word.

Semicolon (;) Use a semicolon to join two closely related independent clauses. Use a semicolon to separate items in a series when the items themselves contain commas.

Square brackets ([ ]) Use square brackets to make additions or comments to quoted material.

Words that are often confused

Some words in scientific writing are frequently confused. Master the following examples and you will be able to use them correctly in the future. If you are not sure of their correct usage, consult a good English dictionary.

accept / except

accept: receive
except: leave out or exclude; but

access / assess / excess

access: the right to use, see or enter
assess: make a judgement, calculate
excess: surplus

advice / advise

advice: (noun) counsel
advise: (verb) give advice

affect / effect

affect: (verb) influence or modify
effect: (verb) produce, cause, bring about; (noun) result

affluent / effluent

affluent: (adjective) rich
effluent: (noun) something that flows out

aggravate / irritate

aggravate: make a bad situation worse, add to
irritate: annoy, vex

a lot / allot

a lot: many
allot: divide, give

allude / elude

allude: mention something indirectly
elude: escape from someone

among / between

among: used with more than two choices
between: used with only two choices

amount / number

amount: indicates something that can be measured
number: indicates something that can be counted

anticipate / expect

anticipate: act in advance
expect: think something will happen

borrow / lend

borrow: use something that belongs to someone else, which must be returned
lend: let someone borrow something that belongs to you, which must be returned to you

cite / sight / site

cite: (verb) quote
sight: (noun) vision
site: (noun) location

compared with / compared to

compared with: look at differences
compared to: look at similarities

complement / compliment

complement: (verb) complete or finish something; (noun) something that completes
compliment: (verb) praise; (noun) praise, free of charge

compose / comprise

compose: make up or create
comprise: consist of, be composed of, be made up of.
Remember, the whole comprises its parts.

confuse / complicate / exacerbate

confuse: bewilder
complicate: make more complex or detailed
exacerbate: make worse

decrease / reduce

decrease: lessen in number
reduce: lessen in amount

dependant / dependent

dependant: (noun) someone who depends on someone else
dependent: (adjective) relying on someone or something else

describe / prescribe / proscribe

describe: give features or details of something
prescribe: set down as a rule or order
proscribe: forbid the practice or use of something

determine / estimate

determine: decide something

estimate: (verb) compute the value of something; (noun) the value or cost of something

discreet / discrete

discreet: tactful

discrete: separate, distinct

do / due

do: (verb) act or make

due: (adjective) caused by, scheduled

elicit / illicit

elicit: (verb) draw out

illicit: (adjective) illegal, not permitted

e.g. / etc. / et al. / i.e. / viz.

e.g.: exempli gratia (for example)

etc.: et cetera (and so forth)

et al.: et alii (and others)

i.e.: id est (that is)

viz.: viz (namely)

farther / further

farther: used with distance

further: used with time or quantity

fewer / less

fewer: used with objects that can counted by number

less: used with objects that can be measured by amount

good / well

good: (adjective) of a high standard, of the right kind

well: (adverb) satisfactorily

imply / infer

imply: suggest or indicate without saying it directly
infer: form an opinion, deduce from evidence

it's / its

it's: it is, it has
its: possessive form of it, belonging to it

lay / lie

lay: place
lie: recline

lead / led

lead: (noun) a soft metal
led: past tense or past participle of to lead

loose / lose

loose: (rhymes with goose) not tight
lose: (rhymes with use) misplace or have something disappear

objective / rationale

objective: goal, aim
rationale: reason, justification

precede / proceed

precede: (verb) go before
proceed: (verb) continue; (noun: proceeds) money

principal / principle

principal: (adjective) main; (noun) person in charge, money
principle: (noun) rule, code

quiet / quite

quiet: not noisy
quite: somewhat, rather

regime / regimen

regime: political organisation
regimen: rules followed

respectfully / respectively

respectfully: with respect
respectively: to each in the order listed

role / roll

role: function, part in a play
roll: (noun) list of members; (verb) move by turning over

stationary / stationery

stationary: not moving, fixed
stationery: paper

that / which

that: used as a defining, or restrictive pronoun
which: used as a non-defining, or non-restrictive pronoun

their / there / they're

their: (possessive) belonging to them
there: in that place
they're: they are

to / too /two

to: (preposition) indicating place, purpose, time, etc.
too: (adverb) also, very, excessively
two: the numeral 2

unique / unusual

unique: sole, only, alone, without equal
unusual: not common

verbal / oral

verbal: using words
oral: spoken, not written

whether / weather / wether

whether: (conjunction) introduces possible alternatives
weather: (noun) state of the atmosphere
wether: (noun) a castrated male sheep

your / you're

your: (possessive) belonging to you
you're: you are

6

LITERATURE SEARCHING AND REFERENCING

Before writing a scientific paper you must read other papers on your topic. Scientific research is not conducted in isolation – it relates to earlier work and either supports or disagrees with it. However, searching through millions of publications can be a daunting prospect. With a bit of planning it can be made much easier.

Literature searching should be performed in the same methodical way as practical research. You should develop a strategy and follow it in a logical way. An important rule is to record where you have searched and what you find. This saves time because you do not repeat searches that found nothing and you can quickly retrace articles that were useful.

SEARCH STRATEGIES

Many people now have access to electronic search facilities, either through Internet search engines or dedicated bibliographic databases in libraries (see examples given under "Further Reading"). The common feature of all electronic methods is the use of *keywords.*

The author of a paper usually specifies keywords, although the administrators of a database may add words. Sometimes keywords are generated automatically by indexing programmes. Keywords indicate the content of the paper and can be used to search for specific subjects. For example, typing "cat" in a keyword search will list all the papers where cat has been chosen as a keyword. Powerful computers allow faster searching, so some database systems allow you to search the title or even the abstract of a paper for specific words. Do not use words like "the" and "of" as keywords – they will usually be ignored.

Keywords are usually combined to form a logical expression. For example, "cat and dog" might be used to find all papers that contained both words "cat" and "dog"; "cat or dog" would find all papers that contained either word. Some search engines use "+" and "/" instead of "and" or "or", so it is important to read the instructions. Words might be excluded by using the word "not" or "-", so the expression "cat not tiger" would list all papers about cats that did not contain the word tiger. Complicated expressions can usually be generated by using brackets, e.g. "(cat not tiger) and (dog or horse)" would list all papers that contained the word "cat" and either "dog" or "horse", unless the paper contained the word "tiger".

You can normally use "wildcards" to truncate keywords and widen their scope. It is not possible on all search engines, but "?"normally signifies "any single letter" and "*" normally signifies "any group of letters". For example, "cat?" means search for cat plus one more letter, which usually covers the plural form (i.e. cats); "cat*" means search for cat with all possible endings (like cats, cattle, catalogue, caterpillar). The "*" should be used with care and should not be used at all for words shorter than four letters.

The secret of efficient searching is to choose your words carefully; if the words are too general, you can get thousands of results; if the words are too specific, you might miss relevant papers. An example search is shown in Table 6:1. The researcher searched a database for references on "The effect of insulin-like growth factors (IGF) on RNA expression in bovine ovarian tissue". Clearly "RNA" is a very general keyword as over 100,000 papers were found (Search 1). Adding the keywords "expression" and "ovarian" narrowed the search (Searches 2 and 3), but only when "bovine" was also included did the number of papers found become manageable (Search 4). When "IGF" was added as a keyword (Search 5), only 5 papers were identified. At this point, you need to consider whether any of the keywords are too specific. If this is the case, you might miss some valuable references because the author has chosen a different keyword. For example, "IGF" may be written in full (Insulin-like growth factors), and expanding the search to "IGF or insulin" adds another six references (Search 6). Similarly, authors may choose "ovary" instead of "ovarian"; allowing for this adds a further six references (Search 7). It could also have been worthwhile to widen the search to "bovine? or cattle or cow?" (or "bovine or bovines or cattle or cow or cows"). Looking at an index or thesaurus might help you to find alternative keywords for your search.

Table 6:1. Results of a database search showing importance of choosing keywords carefully

Search	*Keyword expression*	*Number of papers found*	*Comments*
1	RNA	101,073	Too many papers to examine individually
2	RNA + expression	85,685	
3	RNA + expression + ovarian	1192	
4	RNA + expression + ovarian + bovine	150	
5	RNA + expression + ovarian + bovine + IGF	5	Very specific, may have missed some papers
6	RNA + expression + ovarian + bovine + (IGF or insulin)	11	
7	RNA + expression + (ovarian or ovary) + bovine + (IGF or insulin)	17	

This example illustrates two principles of searching databases. You can narrow your search by adding more specific keywords, and you can widen the search by giving alternative keywords. In this case we started from a wide search and narrowed it down. An alternative strategy would be to start with a very specific search and make it wider. There is no best strategy; only you can decide when you have a reasonable number of relevant references.

MANUAL SEARCHING

A manual search might also be needed

Electronic databases have revolutionised literature searching, but they do have limitations. Because computers deal with precise facts (ultimately, either 0 or 1), programming them to make approximations or associations is difficult. Language is often imprecise, even scientific language, and there are always at least two ways of saying the same thing. The human brain can make associations far more efficiently than any computer, which is why there is still a place for reading the printed word. A computer will find a reference only if a keyword chosen by the author or editor matches a keyword chosen by you. Sometimes you might find a reference that is very relevant to your topic but contains none of your keywords.

There are specialised publications to help with literature searching, such as *abstracting journals* and *contents lists*, although they are being replaced by electronic versions. Abstracting journals gather abstracts from papers published in journals and group them by subject. Looking at a subject area might help you find a useful paper that is described by keywords you had not thought of. Contents lists record titles of papers published in recent issues of selected journals. Again, looking at full titles might help you make an association between your subject area and a particular paper. Keywords can also be used for manual searching.

Review papers are another valuable source of information, particularly when you are researching a new area. Somebody else has already found a list of references for you and interpreted them. Be aware, however, that the author's interpretation may be different from yours, so you should still look at the original papers. Also, the original papers might refer to previous papers that the reviewer chose to omit. Review papers can be found in many journals and conference proceedings. Doctoral theses are also good starting points for a collection of references. Be careful when using sources that have not been subjected to peer review (e.g. web sites and non-refereed journals).

RECORDING YOUR SEARCH

As mentioned before, it is good practice to record your searches to save having to repeat them, and it is essential to record the results so that you know exactly where references come from. Bibliographic software is available for recording references on a computer and many packages allow you to paste information directly from a database search. Alternatively, you can record the information manually on index cards or pieces of paper. An advantage of using computers for your reference recording is that you can easily search among your references, as well as sort them into the correct order and format for the reference list of your paper.

You obviously want to record all relevant papers, but it is also important that you record all available information about each paper. For a journal paper you should record the title, all authors names, year of publication, full journal name, journal volume and part numbers, and first and last page numbers. For papers published in books, you should record the title, all authors names, title of the book, any editors' names, edition number, publisher's name and place of publication. This information will be needed if you refer to the paper in your own published paper.

Other information that can be recorded for each paper is the abstract and keywords, but you might want to assign your own keywords to the paper and write your own extract. Writing an extract that contains a few key sentences is useful because it forces you to analyse the paper. By doing this, you focus on the important message of the paper and decide what information you might later use in your own review.

The ultimate way of recording a paper is to photocopy it or send for a reprint from the author. Photocopying must only be performed within the laws of copyright, which are designed to protect the rights of authors (see section on "Copyright"). Normally you are allowed to photocopy one paper from a journal for research purposes, but the law varies between countries so check before you copy. Remember though that photocopying is not a substitute for reading; a good researcher will still analyse a paper and write down the key messages. When the time comes to write your own review, you can rearrange the key sentences from all your papers, which will provide you with a very good starting point.

REFERENCING PUBLISHED WORK

Your work must be related to what other research workers have already published. This is done through references that acknowledge your sources of information. If you do not acknowledge your sources, you will be accused of plagiarism, which means pretending someone else's work is your own. Hopefully, other people will refer to your work in the future and you will get credit for it.

The exact way in which references are cited varies between journals, so make sure you read the instructions and look at some papers already published in that journal. This lack of standardisation is annoying, but emphasises the need to record full details of each paper in your collection. For example, if a journal requires the first and last page numbers in a reference, and you have only recorded the first page number, you will have to find the reference again.

Citing references within the main text

The two major systems of citing references are by *name* or *number*. When names are used, the authors' names and year of publication are given in the text of your paper and references are listed at the end in alphabetical order of first author's name. When numbers are used, only a number is given in the text and full references are given at the end with a corresponding number. References may be numbered consecutively as they appear in the paper (so the first reference in the text is given the number 1) or they may be sorted into alphabetical order of first author's name in the reference list and numbered in that order (so the first reference in the text may have any number).

To cite someone's work, you make a statement and either cite them afterwards or as part of the statement.

Examples of citing afterwards:

> Experimental evidence suggests that the sky is blue (Smith, 1999). [name system]
>
> Experimental evidence suggests that the sky is blue[5]. [number system]

Examples of citing as part of the statement:

> Smith (1999) observed that the sky was blue. [name system]
>
> Smith observed that the sky was blue[5]. [number system]

With each specific journal you have no choice of system; you have to follow the journal instructions. Some prefer names because it associates the statements with actual people and dates; some prefer numbers because they do not interrupt the flow of text. The name system gives journals far more scope for introducing variations that may confuse authors.

Where a reference cited in the text has more than one author, you will be asked to either list all authors, e.g. Smith, Brown and Yang (1999), or you will be asked to use the abbreviation *et al.*, which is short for the Latin *et alii*, meaning "and others", e.g. Smith *et al.* (1999).

Where an author or group has more than one publication in one year, they are usually distinguished by adding a letter to the date, e.g. (1990a), (1990b), in both the text and the reference list.

Where several references are listed together in the text, they are either in chronological order or alphabetical order, e.g. (Jones, 1998; Black, 2000; Green, 2000) or (Black, 2000; Green, 2000; Jones, 1998).

Writing the reference list

The format of reference lists at the end of your paper is different for virtually every journal. For this reason, you should read the instructions carefully and follow them. The only advice that can be given here is to make sure your references are correct and complete. Every reference made in the text should have an entry in the reference list and all references in the list should have been referred to in the paper. An important purpose of the reference list is for others to be able to find the papers you are referring to; incorrect or incomplete references are useless, which is why you need to record all details during literature searches.

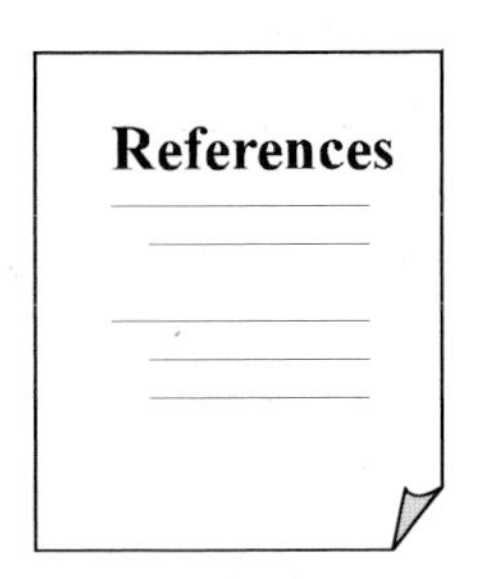

If an author's work has not been published, but you still want to refer to it, you do not include it in the list of references. In the text, you use a form such as Smith (2000, personal communication), where the author has given you some information (and permission to quote it!) or Brown (unpublished data) where the information is not yet in print. Note that many scientific journals do not allow you to use unpublished data.

REFERENCING WEB ADDRESSES

The World-Wide Web is becoming increasingly important as a source of information. Remember, however, that most web sites are not refereed or checked for accuracy, the contents may change at any time and the web page may disappear without notice. For this reason, web addresses are not acceptable in scientific journals, although they may be used in popular articles. Because you are using someone else's work, you must still acknowledge the source. There are no firm rules, but the following guidelines should be followed in the absence of other instructions:

Within main text

Quote the full address, usually starting with **http://** and then giving the **complete** names of all sub-pages, follow this with the full date on which you last accessed the location.

Example:

> These results follow the general decrease in infant mortality rate in the UK between 1965 and 1995 (http://www.unece.org/stats/trend/gbr.htm, 30/06/00).

Within Reference list

You should ensure the following information is available:

- The name and any professional affiliation of any individual associated with the data/page
- The name of the organisation hosting the site
- The title of the page
- The full date on which you last accessed the site
- Any source date attached to the information
- The full web address

Example:

> USDA (1998) United States Department of Agriculture Office of Communications. *Agriculture Fact Book 1998, Chapter 1: U.S. Agriculture-Linking Consumers and Producers, Section 1: What Do Americans Eat?* http://www.usda.gov/news/pubs/fbook98/ch1a.htm (Accessed 30/06/00).

COPYRIGHT

Copyright exists to protect the rights of authors to benefit from their work; more specifically, copyright is designed to prevent others from benefiting from an author's work without permission, for example, by passing the work off as their own, by avoiding payment when photocopying, by mounting on a web site without acknowledgement. Infringing copyright carries serious penalties, so make sure you always cite your sources of information and seek permission to use illustrations or substantial quotations.

Copyright does not protect an idea; it protects the form of literary and artistic work. In the case of scientific papers, this means the text, tables and diagrams, but not the title. The important word is *form*; you are allowed to repeat what someone has written if you change the wording, for example. With diagrams, however, you must make significant changes to avoid infringing copyright.

Copyright is covered by different laws in different countries, but the principles are the same; you cannot reproduce anything in an unmodified form without permission. Protection lasts for 70 years after the author's death, or 95 years from publication of an anonymous work, but it is wise to assume that all printed material is covered by copyright. Copyright is established as soon as original material is created in a form that could be copied. A copyright notice can be placed on the work (e.g. © 2000 Mary Smith) to identify the copyright owner, but this is not a legal requirement. The copyright notice is often only found on one of the front pages of a book or journal, but it applies to everything in the publication.

Photocopying is generally prohibited by copyright laws, except that you can make one copy for research or private study. Do not assume that articles found on web sites are free from copyright; the web author either owns the copyright, or may not have obtained permission to mount the material.

As an author of a scientific paper, you need to be aware of copyright restrictions. The US situation is described on the Library of Congress web site (http://www.loc.gov/copyright); the UK law is described by the Copyright Licensing Agency web site (http://www.cla.co.uk), which also has links to sites in many other countries. In most copyright laws there is reference to "fair dealing", which allows you to reproduce material for research or private study (e.g. photocopying as above) or for the purposes of criticism and review, without obtaining permission. The laws do not state how much you can reproduce in a review, but passages of up to 400 words from a book or 5% of an article, are often accepted as upper limits. Unfortunately, only the Courts can decide whether you break copyright limits, so it is best to seek permission from the copyright holder if you use more than one or two sentences. Permission is not needed if you prepare a table or diagram from data published elsewhere, for example, combining data from several papers; drawing a graph of data published in a table. In all cases, you should acknowledge the source with a proper scientific reference (see section on "Referencing").

The copyright holder is usually the author of the work, but copyright can be transferred, or a licence granted, to their publisher; this will be shown in a copyright declaration. If an author produces a literary or artistic work as part of their employment, e.g. a journalist or writer of instruction manuals, copyright usually belongs to their employer. If you want to reproduce substantial text, a table or a diagram from someone else's paper, ask for permission. You should normally write to the publisher stating what you want to reproduce (full reference), where you want to reproduce it (full provisional reference) and whether you intend to modify the material (if you want to change a diagram, for instance, enclose a copy of your proposed version). The publisher may charge a small fee for permission to reproduce the material; if the fee is too large, either seek advice from your own publisher or omit the material. The following is a sample permission form:

"Dear Sir or Madam,

I am writing a [paper / chapter / book] provisionally entitled [Title], which will [be submitted to the journal *Journal* / be published by *Publisher*]. I would like permission to reproduce the following material from the paper [*Full reference*].

Figure X, Table Y

I would like to change the regression line in Figure X to a solid line because I feel this will be clearer than the current dotted line.

Full acknowledgement will be given to the source of the material; if you require a particular wording, please let me know.

Thank you for your assistance. I look forward to hearing from you.

Yours faithfully,"

Larger publishers have a permissions department, so address your letter there. If the publisher refuses permission, seek advice from your own publisher. You may be able to modify the material sufficiently to be outside the original author's copyright. Under no circumstances should you use the original material without permission.

7

GETTING A PAPER INTO PRINT

Research that yields new knowledge should be written up as a scientific paper and submitted for publication to an appropriate peer-reviewed, scientific journal. The peer-review process is essential for maintaining scientific standards and ensuring that your paper is worthy of publication. In addition, editors check the quality of scientific writing and clarity in presentation of concepts. Submission of the paper to more than one journal is considered to be a breach of professional ethics. A journal will not knowingly accept a paper that has been, or is scheduled to be, published elsewhere in the scientific literature; a published abstract, at a conference for example, is an exception. Duplicate publication is also a violation of professional ethics.

Choose a journal that publishes papers in your field. The choice of journal also depends on who you want your audience to be ... who do you want to read your paper? That may or may not be where other papers in your field are being published ... it often is, of course. You should know which journals are relevant from your review of literature, but your choice may be a limited if your paper is very specialised. If there are several options, go for a journal that is widely read. Some institutions encourage the use of journals with a high impact factor. Impact factors are calculated from the number of citations received for papers in a journal (see http://jcrweb.com); this is a measure of visibility and does not indicate quality, since a bad paper can be cited many times for being bad and the calculation can include self citations. Visibility, in itself, is important because publishing in a journal that few people read does not help your research.

PREPARING YOUR MANUSCRIPT FOR SUBMISSION

Consult the journal for the details (e.g. the number of copies), the form of submission (e.g. paper or electronic), and where to submit the paper. Make sure you follow instructions about page layout (margins etc.), line spacing (usually double), line numbering (usually required), referencing and format of tables and figures. Although all journals will edit accepted papers for style, clarity of writing and grammatical correctness, they may reject your paper if you have not followed their instructions on typesetting.

Some journals require a non-refundable processing fee for each paper submitted. Many journals ask authors to complete and submit a "Manuscript Submission and Copyright Release Form", which among other things assures that each of the authors has read the paper and that the paper is not being submitted for publication elsewhere, and releases copyright of the paper to the journal. Copyright is discussed in the chapter "Literature searching and referencing". You do not *have* to give copyright to a journal but can grant the journal a licence to use your material; in some instances you may not be *allowed* to release copyright – if it is not yours to release because it belongs to your employer, or, if you are a government employee, because it belongs to the public.

AUTHORSHIP AND ADDRESSES

There is great debate about authorship of papers, including the order of authors and who should be listed. These matters should ideally be agreed at the start of the research programme, but decisions are often postponed until the paper is nearing completion, so we are highlighting potential problems here.

The decision on order appears easy when only two people have contributed to the work; the person who does the research and writes the paper usually comes first and the other comes second (e.g. student then supervisor); if the second author substantially rewrote the paper, the order might be reversed. It is trickier when the paper reports work from a team. Research groups have different conventions; alphabetical order, order of contribution (greatest contribution first), reverse order of seniority (Professor last, although note that the first author is referred to as the "Senior Author"). If the team is from more than one institution, authors may be grouped by address. The order is immaterial, except that the first author of a long list will have the honour of being referred to with "*et al.*". One author might like to be the last author on a list because it implies seniority in some systems, but this is offset by other systems where the last author made the least contribution.

The list of authors should include only those who made a substantial contribution to the research work and/or the writing of the paper. Do not include your Head of Department unless he or she was actively involved; the old practice of putting the Head's name on every paper was just a waste of journal space. Similarly, do not put a colleague's name on a paper in return for getting your name on theirs; this dilutes the credit given to the real researchers. A good scientist will not ask for their name to be included, nor include anyone else's name, unless they have been closely involved with the work. On the other hand, make sure that everyone who has made a substantial contribution *does* get listed as an author. The contribution should normally be intellectual, e.g. original concept, obtaining funding through a grant application, designing experiments, supervising the project, writing the paper. Lesser contributions should be listed in the acknowledgements, even a technician who did all the experimental work if they were just following instructions from the scientist. All authors jointly share responsibility for the contents of a paper, so make sure that all authors have read the final version before submission.

Addresses of authors should be given in the journal style. Make sure the full international address is given so that interested people can contact the authors. The first address should be the institution where the majority of the work was conducted (normally that of the first author). If the first author has relocated after completing the work, the institution address should remain the original one and a footnote should indicate his or her current address. With more than one institution, superscripts are often used to indicate who works where. One author may be identified for correspondence and reprint requests.

SUBMISSION

Double-check that you have followed all the instructions and send your manuscript to the correct address shown in the journal. Enclose a covering letter stating your name and full contact details (postal address, including ZIP or postal code and country; telephone; fax; email address), together with the title and authors of your paper. Use a strong envelope or package to avoid damage in the post. You

You may have to wait a few weeks

should receive an acknowledgement of receipt within a few days. You will then have to wait a few weeks to hear if your paper has been accepted. During this time, your manuscript will be carefully considered by an editor and one or more (usually two) referees; this is the Peer Review process.

EDITOR'S AND REFEREES' REPORTS

You will receive reports from the editor and the referees. Each referee's report will give opinions on your paper and will normally make suggestions for improvement. Often the editor's report simply consists of a covering letter that interprets the report from the referee(s), but it will contain a decision about whether your paper is acceptable for publication.

Most journals have four possible decisions:

1. accept the paper as it is written
2. accept the paper after minor revision
3. accept the paper after major rewriting
4. reject the paper for publication

Decision 1 is very unusual (less than 5%), particularly for inexperienced authors. Most referees and editors will recommend at least some small changes, if only to prove that they have thoroughly assessed the paper.

Decision 2 is normal for a paper that describes sound science and is well written. Minor corrections can usually be dealt with quickly and the manuscript returned to the editor for publication. Make sure you read the recommendations carefully and follow them if they are acceptable. However, do not be afraid to question any with which you do not agree. Referees and editors can make mistakes and may not have understood something you wrote. If this is the case, however, you should still consider rewriting the relevant section so that your meaning is clear and unambiguous. Some journals require

Papers are rarely accepted without changes

a covering letter with the revised manuscript detailing how you have addressed each point in the referee's report. It is good practice to include this, even if not specifically requested.

Decision 3 is similar to *Decision 2*, except the changes requested will be either bigger or more numerous. Again, read through the reports carefully and see if you agree with them. Usually changes are suggested that greatly improve your paper and you should follow them carefully. If there are any that you do not understand, or disagree with, ask the editor for clarification. Editors welcome requests for help if it saves them having to return the paper to you again at a later stage. Always make your queries clear and short. You do not need to repeat large sections of the paper or report, since the editor will have copies. Examples of major revision include recalculating data, correcting statistical analysis, adding more detail to materials and methods, and lengthening or shortening the discussion.

Decision 4 can be a shock. You have spent a long time gathering data and preparing a paper, only to have it rejected by anonymous referees to whom you cannot complain. You may feel angry or disheartened, both of which are natural feelings. When you have recovered, examine the reasons for rejection – it could be that the decision might be reversed after major revision. Has the referee misunderstood something in your paper? Could you change the emphasis of the paper to make it acceptable? Is the subject material more appropriate for a different journal? If you have a valid reason for appeal, write to the editor, but first you ought to check your arguments with an experienced colleague. If the paper was rejected because of flaws in experimental design or low scientific merit (assuming you described the experiment accurately) there is nothing that can be done to recover the situation. You should accept the decision gracefully and learn from your mistakes. Do not waste other people's time by submitting to another journal. If a less discerning publication does accept your paper with fundamental flaws, it will not do your reputation any good and you will be embarrassed to read it in later years.

AUTHOR PROOFS

Once accepted by the editor, your paper is forwarded to a Technical Editor to prepare it for publication. The Technical Editor looks in detail at the way your manuscript is written, rather than its academic merit. They may contact you for missing information (e.g. incomplete references), but normally the paper is typeset, figures are reproduced, and author proofs are prepared.

Author proofs, together with the manuscript, are sent to the corresponding author indicated on the title page of the paper. To make corrections to the proof for English-language journals, standard proof-readers' marks are available, as published in Hart's Rules[1] or BS 5261C[2]. They can be obtained from the journal and may accompany the proofs, or can be found in some English dictionaries. Author proofs must be dealt with promptly, so as not to delay publication. You will probably be reminded that only typesetting and factual errors should be corrected at this stage; if you try to introduce new material you may have to pay. One new line might mean that all subsequent pages in the journal or book have to be typeset again, which is very expensive.

Before you start reading the proofs, make a copy and work on the copy, rather than the original; mistakes in your proof-readers' marks can then be changed without causing confusion for the printer. Firstly, read your proof through quickly or, even better, get someone else to read it, while you follow the manuscript. This will show you if any words have been left out, or new words inserted. Secondly, go through the proof one word at a time, checking the spelling. Professional proof-readers sometimes start at the end and read the proof backwards; this is because your brain is very good at subconsciously correcting errors as you read. Finally, check every number in your proof, particularly in tables, against the original manuscript; you are the only person who can verify the data, since incorrect numbers will still look "correct" to other people. Corrections to the proof should be made neatly and clearly in the margins of the proof. When you are satisfied that you have identified all the corrections needed, copy the proof marks onto the original proofs and return them with the manuscript to the Technical Editor.

An example of a corrected proof is shown in Figure 7:1; note that proofs do not normally contain so many errors:

---

[1] *Hart's Rules for Compositors and Readers at the University Press Oxford.* Oxford University Press, Oxford.

[2] BS 5261C : 1976 Marks for Copy Preparation and Proof Correction. British Standards Institution, 2 Park Street, London W1A 2BS.

The ABC of Science Communication

Communicating science usually means communicating new knowledge or summarising the present state of knowledge. It is important for the audience to catch the message with as little misunderstanding as possible and to feel confidence in what is written or said.

The ABC of science communication is that it should be:

- Accurate and Audience adapted
- Brief
- Clear.

Science is international. This means that most of those who read listen to a scientific presentation will be doing so in a foreign language. This further emphasizes the need for clarity and for the presentation to be logical, consistent and coherent. Communication is a two-way process. Information cannot merely be delivered - it must be well received and understood as well. The message delivered may be accurate, brief and clear, but yet not be received and understood. This may happen if what you write or say does not relate to the frames of reference of audience. Adapting to the audience, therefore is very important.

A basis for the process scientific is to formulate a hypothesis, which means that you pose a question and a hypothetical ANSWER. Questions and answers are the basis for communication as well. For effective communication you cannot just think of your own topic and the message you want to deliver. You must also consider what questions your audience might have with regard to your topic. Some components of effective communication are indicated in Figure1:1.

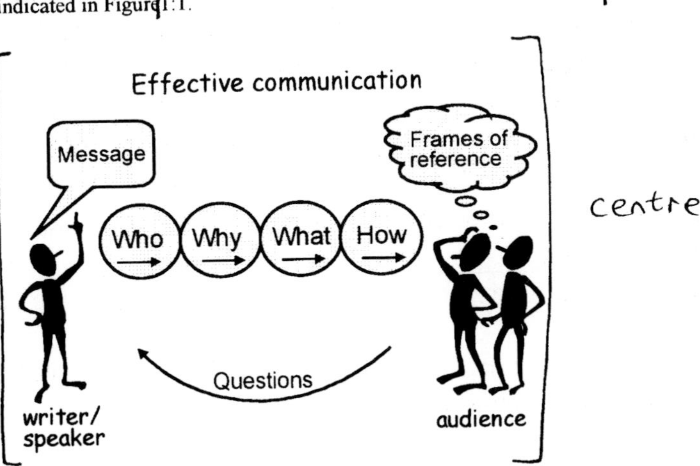

Figure 11. Some components of effective communication.

Figure 7:1. Example of a corrected proof.

The corrections are black on the printed page, but you should write them in red or blue. Table 7:1 shows examples of common proof marks that may be used to correct a manuscript. If you do not feel comfortable using these marks, underline or circle the text and write your correction in the margin, clearly and in simple English. If you write instructions in the margin, draw a line around them so that they are not confused with text to be inserted or changed. For example, if you write "align" in the margin, you might mean "align columns of text" or "insert the word align". Your corresponding mark in the text should make your intention clear, but if "align" is circled it is obviously an instruction. You could make instructions even more distinctive by writing them in a different colour.

Table 7:1. Example proof marks (based mostly on BS 5261C:1976)

Mark in text	*Mark in margin*	*Meaning*
⋏	a ⋏	Insert character (***a***)
⋏	new ⋏	Insert word (*new*)
⋏	◇A ⋏	Insert extra text provided on a separate sheet marked A
/	∂	Delete character
⊢⊣	∂	Delete word(s)
⁀/‿	⁀∂‿	Delete character and close up
⁀⊢⊣‿	⁀∂‿	Delete word(s) and close up
/	Correct character	Replace character
⊢⊣	Correct word	Replace word(s)
⁀‿	⁀‿	Close up (less space)
Y	Y	Insert space
//	align	Align columns

Table 7:1 Continued

Mark in text	*Mark in margin*	*Meaning*
═	*align*	Align characters in a row
⌐	*new para*	Start new paragraph
⌒	*run on*	No new paragraph
⊔⊓	⊔⊓	Swap characters or words
2 3 1	1 2 3	Change order of words
⊢ ⊐	⊏	Move words left
⊏ ⊐⊣	⊐	Move words right
▭⟶	*move*	Move words to given position
[ ]	*centre*	Centre
⋏ or /		*Insert or delete:*
	⊙	Full stop or decimal point
	(,) (;)	Comma, semi-colon
	‘’	Quotation marks
	(/)	Slash
______	*italics*	Change to italics
~~	*bold*	Change to bold
______	*l.c.*	Change to lower case
______	*CAPS*	Change to upper case (capitals)
	/	Mark between corrections in same line
-------------	*stet.*	Leave unchanged (mistake in proof-reading)

If you have two or more corrections for one line of text, separate each correction in the margin by a slash (/). You can use both margins, but the order must be left to right across both margins, in the same order as the line of text. If there is not enough room in the margins, place a correction at the top or bottom of the page (clearly indicated), or on a separate sheet of paper. Never write on the back of the proof or your corrections may be missed. Do not enter corrections between lines of text; they are hard to read. Do not use capital letters when writing corrections (unless the material should be in capitals); the typesetter might set the corrections in capitals.

Some journals charge authors for each printed page, to help cover the cost of publication. A form for ordering reprints is forwarded to the corresponding author with the author proof. The form normally indicates the page charge and the charge if any for reprints; make sure you return this form with your proofs.

8

ORAL PRESENTATION AND VISUAL DISPLAYS

New research results, as well as updated reviews of specific research areas, are communicated at scientific meetings, such as national and international conferences, often even before the results are published in scientific journals. This communication usually occurs in the form of a conference paper (sometimes just an abstract), plus an oral presentation or a poster. Oral presentations of research results are also given frequently at other types of meetings, for instance at workshops and university seminars, as well as at public meetings. In this chapter we will mainly focus on oral presentation at scientific meetings and seminars, but what is said is also largely valid in other situations.

Successful presentations can be given in many different ways, as long as your message is conveyed effectively. In fact, variability between presenters adds to the quality of a meeting. Choose a form that fits your personality, but make sure to adapt to the audience, to speak to be heard, to use only visuals that can be seen, and to show your own interest in the topic!

PLANNING THE ORAL PRESENTATION

It is worthwhile making a real effort to do good oral presentations; not only because audiences usually remember both the good and the bad presentations, but also because the message is carried through most efficiently in the good ones.

Considering the amount of work that goes into research, you should make the most of every opportunity to communicate the results and conclusions, to spread knowledge and stimulate action, and to get valuable contacts.

You have a responsibility to your audience; their time is precious. A presentation should add something to just reading your paper; it should arouse interest in the topic and stimulate discussion, as well as making it easy for the audience to quickly catch the main points.

Who – Why – What – How?

The reason for your scientific presentation at a meeting is usually that you have new research results to communicate, or that you have been asked to review a topic. What to present depends on the purpose of your presentation, what messages you find interesting and important to communicate, and who will be your audience (see the chapter "Communicating Science"). Adapting to the audience is even more important in oral than in written communication. The listeners have to catch what you say immediately; they cannot go back and check again as they can do in written text. So, think ahead about your audience; what are their backgrounds, interests and capabilities? Your speech will not be understood and retained unless it relates to the frames of reference of the audience. You often find that the audience for your oral presentation might be quite diverse, even at scientific meetings, and that your presentation needs to be at a more general level than your written paper. At the same time, though, you shouldn't lose in-depth coverage of essential parts.

It is sometimes pointed out that the word "presentation" contains the word "present". What you do is to give the audience a present, something you like to give and hope they will enjoy. Your gift has value, which is enhanced by you being present at delivery.

Some features of scientific meetings

Scientific meetings vary from national meetings, where most participants speak their mother tongue, to international meetings, where many of the participants must speak and listen to a foreign language. Some meetings concentrate on a specific topic, whereas others cover a larger area, often split into a number of parallel sessions. In those meetings the audience is often mixed with regard to research specialisation, and not everyone has a specific interest in every paper. Usually the written papers are available at the presentation, but they are often not read in advance by the audience. Sometimes only an abstract is provided.

The audience is often large, and activity in the form of questions/discussion generally takes place at the end of each presentation or session. The time schedule is tight, because there are normally a large number of oral presentations in a day. The presenter of an invited paper might get 20–30 minutes, whereas the presentation of a scientific research paper is often restricted to 10 minutes. This further emphasises the importance of oral presentations being attractive, interesting, easy to understand and to focus on a main message. Listening to presentations for many hours per day, often in low lighting conditions, and in a foreign language as well, is quite hard for the audience.

PREPARING THE ORAL PRESENTATION

When preparing an oral presentation you need to think of how you will best raise and maintain the audience's interest, and how to structure the topic so that the audience will understand and remember your main messages.

Audience attention varies during a presentation

As soon as a presentation stops being interesting enough, the thoughts of your listeners will easily slip away to something else. This means that you, as a speaker, must be more interesting than the listener's own thoughts for a message to be taken in. Disturbing factors in the room can also cause parts of your presentation to be missed by the audience. There might, for example, be noise from the projector or from people moving around between sessions. Audience attention is mostly at the highest level at the beginning and end of a presentation (Figure 8:1).

At the beginning of your presentation, audience attention is high due to curiosity about what will come, and at the end it rises again as conclusions are expected. In the middle of your talk, however, when attention is the lowest, some variation

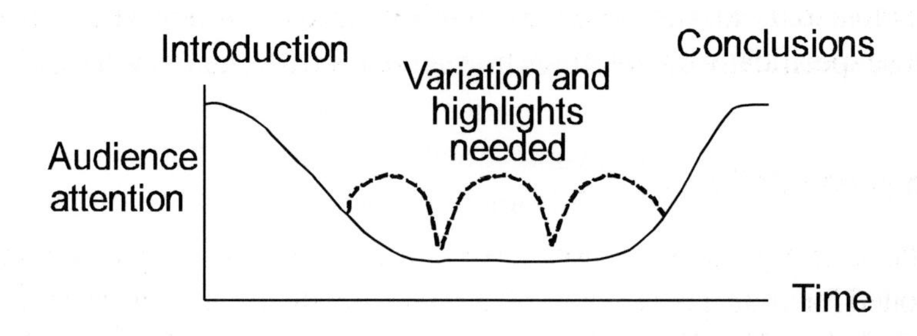

Figure 8:1. Audience attention varies during a presentation.

and highlights are needed to maintain interest. Because the audience can easily miss or mishear parts of a presentation, the most important points might need to be emphasised twice to give the listeners a reasonable chance to catch them. Repetition, if possible with some variety, increases the transfer of your message. Plan your speech so that you use the times when audience attention is at the highest level to get your main messages through.

Anticipate audience questions

Having considered what you want to achieve with your oral presentation, you should also consider the needs of your audience in more detail. One way to do this could be to write down a number of questions that you anticipate your audience might have with regard to your topic, and incorporate your answers to some of these in your presentation.

Formulating probable audience questions will:

- make your presentation better adapted to the audience
- give your motivation
- be helpful in the discussion following your presentation

Many of the audience will probably be more interested in the significance and possible implications of your results than in the research *per se*. They want to know how the results apply to them. The specialists, on the other hand, may be more interested in the methods used. If the audience is mixed, you need to do your best to satisfy all the needs, but don't go into too much detail. You may invite those specifically interested to discuss details with you individually later on.

Make a presentation outline

Before developing a presentation outline, you must know how much time has been allotted for your speech. Most of us overestimate the amount of information to be included, and it will save work later if you do some realistic planning with regard to presentation time from the very beginning.

It is also wise to consider the language for your presentation. Will it be in a language other than your mother tongue? In that case, prepare in the foreign language directly, if possible. Doing so will make it easier to start thinking in that language, and you won't get stuck with words or sentences in your native language that may be difficult to translate. If you are fortunate enough to be allowed to speak in your mother tongue, remember that for a large part of the audience this may be a foreign language, so plan for simplicity and clarity.

How to outline a presentation cannot be given a single answer; that depends on the subject, the audience and your personality. The same structuring principles that were discussed for written communication, however, can be applied in an oral presentation (see the chapter "Getting Started in Writing"). You might choose order with regard to chronological events, interest/importance, cause and effect, or "pros and cons".

The chronological structure resembles the scientific writing model. Using that for a presentation outline probably means starting with the objectives and historical background, then presenting materials and methods, thereafter results followed by a discussion, and finally the conclusions. This model very much relates to your research work, but may not fully connect with the needs of the audience.

General structure for a presentation

You may not want to strictly follow a single structuring principle, but may prefer to make a synthesis of several. A general structure for a presentation outline could be:

- Title
- Introduction
- Body
- Summary/conclusions

Title. Expend effort on making a good title for your presentation and the corresponding paper. The title is given in the meeting programme, and should encourage people to come to your presentation, and to read your paper as well. Try to make the title both informative and inviting. It should be as direct as possible and not too long. Don't be afraid to state a problem or pose a question in the title, or even make it provocative; that usually attracts the audience. Also remember to acknowledge your co-workers at the start or end of your presentation.

Introduction. In the introduction you should take advantage of the higher level of audience attention at the very beginning of a speech to make the audience really interested in listening to the rest of your presentation. Give the context, background and motivation for the topic you will present. Make the audience understand that your message is important to them. One idea could be to start from some of your postulated audience questions. It is sometimes recommended that you should also briefly outline your presentation plan. Be careful, though, not to do this in too much detail, or your listeners may lose their desire to listen further. The introduction must not be too long.

Body. The body constitutes the major part of your presentation. As the audience attention may drop during this part, it is important to structure the content so that it is both logical and interesting, and so that it relates to the audience. Use analogies, if needed, to explain things better, and to answer some more of your postulated audience questions. Interpretation and analysis of facts presented, comments upon problems and prospects, advantages/disadvantages of various solutions, and some sub-conclusions, are all aspects that can be vital to include in the body. Furthermore, try to give realistic examples in connection with theoretical issues.

Divide the body into sub-sections, and structure each in the way you find the best. If the audience is mixed with regard to specialisation, it is probably better just to give a brief outline of the methods used, and their place in the body can vary. Applying the chronological model means that the methods will be presented before the results, whereas in the interest/importance approach it might be the opposite.

To maintain audience interest during the body part of your presentation, it is essential to build variation into it. Using visual aids, body language and voice modulation are some alternatives. Another option could be to allow and to stimulate questions from the audience during your presentation. This may be difficult to handle in scientific meetings where the audience is often large, but it can be well suited for workshops and seminars.

Summary/conclusions. This is the most important part of the presentation, because what you say here is what the audience will normally remember most. So, make use of this opportunity to really get your main messages through! Summarise your main findings and conclusions for their interpretation and significance, and your possible recommendations.

Make sure you get to the summary/conclusion part in your presentation before your time runs out. Remember, however, that audience attention will drop again if this part is not short. Therefore, keep the summary statements concise and to the point, and limit the number of statements. Having delivered the final conclusions, end your presentation with just a few words to indicate that it is completed, and stimulate questions/discussion if applicable.

VISUALS SUPPORT YOUR SPEECH

Supporting a speech with visual displays is very useful in science communication, as well as in providing variation and in stimulating interest throughout the presentation. Tests have shown that people absorb more, and retain the information better, when it is communicated to them in both words and pictures, as compared with words only (Figure 8:2).

It is essential that the visuals used are clearly seen and meaningful to the audience. They should relate to the words spoken, be well organised and emphasise the important points. A visual that is overloaded, difficult to read or understand, or not apparently related to the speech will only be distracting.

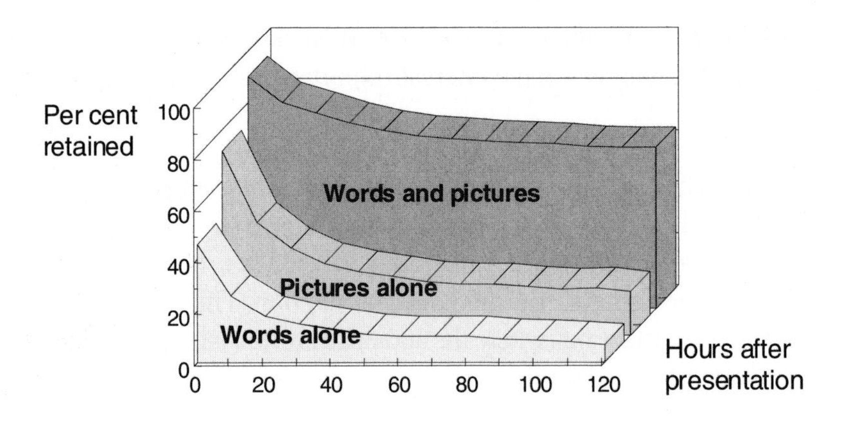

Figure 8:2. Audience retention of presented information[1].

Which visuals to use?

The visuals mainly used in oral presentations at scientific meetings are:

- Overhead transparencies (acetate sheets)
- Slides for a projector
- Electronic presentation (computer-based projections)

Other visual options are films and videos. In a small meeting it might also be an option to use flip charts or a writing board.

When deciding on which type of visuals to use, you need to find out what technical equipment will be available in the presentation room. Another factor that might limit your choice is the equipment on hand to produce your visuals. Remember though, that using the most modern methods for producing and showing visuals is no guarantee of a successful oral presentation. In fact, technical sophistication may even be a drawback, because there is more that can go wrong, and it might require quite a lot of practice and good luck. You can do a lot with simple means. For example, if you don't have access to a colour printer, you can still add colour to transparencies or make a drawing by hand.

---

1 Modified from Woelfle, R.M. (Ed.): A guide for better technical presentations, p.37. IEEE Press:New York (1975). Results from US Air Force.

Keep room lights on as much as possible

Illumination of the presentation room is an important factor to consider when producing visuals. If the room light is on during your presentation, you can glance at your notes without being tied to a lectern. It will also be easier to keep a more natural contact with the audience and to notice whether people can follow you.

With the room illuminated during your speech the audience will:

- see you better and experience more contact
- see to make notes
- not get tired so easily.

So, think of the room light when you plan your visuals!

The room light can usually be fully on when transparencies are shown. Using slides, however, often requires the room light to be reduced or switched off, especially when the slides shown have a dark background. Electronic presentation may require reduced room light, unless powerful modern equipment is available. You should use colours in your visuals that minimise the need to reduce illumination, and tell the person handling the light not to darken the room more than necessary. Also make sure that full light is turned back on as soon as visuals are no longer being used.

Using an overhead projector

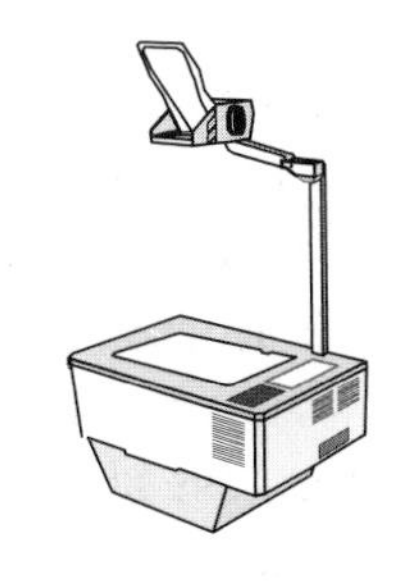

Using overhead transparencies in a presentation is usually reliable. To operate the overhead projector the speaker might leave the lectern, which can make the presentation alive and varied, especially as the room can normally be illuminated. It is also an advantage that you can change the order of, or exclude/ add transparencies at a late stage, or even during the presentation without the audience noticing.

One problem in using transparencies is that the overhead projector is not always placed where it is most suitable for the speaker to stand; it is sometimes not even placed on the stage. In such situations, you may need someone assisting you in switching transparencies. It should be obvious to organisers that the overhead projector must be placed where it is convenient for the speaker to use it (without standing in the way of the viewers), and that you need a surface nearby for your pile of transparencies. Moreover, the projection screen must have the right angle and be large enough for the size of the audience.

An overhead presentation can be improved if you think of the following:

- Don't block the view for the audience. Step aside a little. If you need to point at something directly on the transparency, take care that your body doesn't cover what is projected on the screen. If arrangements allow, it might be better to point at the wall screen, but make sure you do not turn your back on the audience. You need to face them, so use your arm nearest the screen to point with.
- Try to avoid using a sheet of paper to release the transparency content in stages. Audiences often don't like it, and you get tied to the projector.
- Transparencies are usually better focused if not shown in a plastic folder, and the string of holes is disturbing, unless you hide it. If you intend to take the transparencies out of the folders, do it before your speech starts. Speakers sometimes place a paper between each transparency, but be aware that this may cause both trouble and noise.
- Cover the glass surface of the projector when you aren't using it in parts of your presentation, to avoid the audience being distracted by a light screen on the wall. You may also cover the part of the surface not occupied by the transparency.

Using a slide projector

Slides are often chosen when photos are to be used as visuals, but text, tables, graphs and other illustrations can also be shown as slides. Using slides instead of transparencies means that the speaker doesn't need to bother about switching visuals manually, as this is done with a remote control, or by a person handling the projector. You should check beforehand that your slides will be shown in the right order and orientation.

A major disadvantage of using slides is that the room light normally needs to be reduced, and that the speaker usually gets tied to the lectern, where the only light is. Slide projectors are sometimes quite noisy, which may be a disturbing factor. Another disadvantage is that you cannot always exclude a visual from your presentation without the audience noticing, and it may be difficult to go back and show a certain slide again if needed in the discussion.

For the preparation of slides, be aware that it may take several days to get them produced photographically from your computer displays. Pay attention to the fact that the standard thickness for slide frames varies. A European standard frame (3 mm) may get stuck in a projector magazine in the US, where the standard is 2 mm; frames thinner than 2 mm, e.g. paper frames, sometimes fall straight through and block the projector. Therefore, find out what standard to use before you assemble your slides!

Electronic presentation

Your visuals can be shown effectively by using presentation software (such as PowerPoint, Persuasion) and projection equipment. Before planning for an electronic presentation, make sure that the presentation room has the equipment needed and that it is compatible with the software and computer/disks that you use. Moreover, learn how to solve compatibility problems.

An electronic presentation offers several possibilities, such as:

- smooth transition from one display to another
- sequential build up of bullet points
- animation of graphs and figures
- hidden displays in readiness if needed
- flexibility in where to stand if a remote control is used

Presentation software provides you with lots of opportunities, but it is best to use it with discrimination. Remember that the focus should be on your content. The tools used should help to emphasise your message, not draw the audience's attention to special effects.

One option that might be useful is the "build effect", particularly with displays containing bullet points. You can thus reveal them one at a time, and don't need to use a pointer. Previous bullets in the same display can be set to fade slightly, but it is best not to make them disappear. Avoid, however, having lots of movement within each display. A text slide, as well as a graph, can be difficult to grasp if it is changing all the time.

A factor to consider in relation to electronic presentation is the room illumination. It should normally be possible to keep room lights on, but you may need to choose colours that give the best conditions for this, and that is usually dark text on a bright background. Also, be aware that colours may look different in your presentation, compared with what they looked like on your computer screen.

As there are many things that might go wrong in an electronic presentation, it is wise to be prepared with back-up displays in the form of ordinary transparencies or slides, especially if your presentation will be abroad at an international conference.

USE A COMPUTER TO CREATE YOUR VISUALS

All the types of visuals discussed above, i.e. transparencies, slides and electronic displays, can be created on a computer, and the same display can be used for all three purposes. For producing the visuals, you can use software for presentation, word processing, graphics and drawings. You can enter clip-art and photos into your visual displays, as well as visual information from other sources, e.g. the Internet. Make sure, though, that you are not infringing copyright.

Choose colours and background that emphasise the message

Visual displays can be produced on a computer in various ways. You can use autolayouts and various background templates that help you build displays in a standard format, or you can start from a blank page and give them a more personal design.

Presentation templates (from which you normally choose one) aim at harmonising your visuals, so they will all have the same background colour and pattern, such as a spiral, ribbons, spots etc. Whether this is the best way or not isn't easy to say. It's a matter of taste. Having the same background appearance for all visuals in a presentation might seem professional, but it can also be boring, especially if the previous speaker happened to choose the same template as you. Showing a visual that is not too similar to the previous one can be an efficient way of capturing audience attention, but you should still strive to be consistent in the use of font(s), font sizes and colours. Remember that the background used in displays must not be distracting; if it is, it would be better with a plain background.

Colour can make visuals more attractive, but using too many colours in a single display may distract the viewer from the message. Try to:

- Use colours that enhance the message, e.g. contrasting, harmonising etc.
- Be careful in basing a distinction solely on red vs. green. Those in the audience who are colour-blind will miss it.
- Make a good contrast between text and background, i.e. dark text on a bright background, or the opposite.
- Choose colours so that the room can be illuminated as much as possible.

Clip-art (prefabricated pictures) can be very useful for making attractive visuals, assuming that they support and emphasise the text message; otherwise they may just distract from it. To get a desired illustration, you can ungroup and make changes in a clip-art or combine different ones, but first check any restrictions that came with the clip-art.

Some examples of visuals (made in PowerPoint) with different layouts and backgrounds are shown in Figure 8:3. Figures 8:3a-c were produced using a presentation template (provided with the software) for the background and an autolayout for the content, whereas Figure 8:3d was composed by writing the text on a blank page with no fixed layout, and inserting clip-art.

Note that the displays in Figure 8:3 are done in ***horizontal format*** (landscape orientation). This is recommended, as the projected image then can be positioned high on the wall screen for better viewing. It also allows longer lines of text, so that fewer lines are required to complete a point. This makes reading easier at a distance. Make sure, though, that your displays don't exceed the width of the screen.

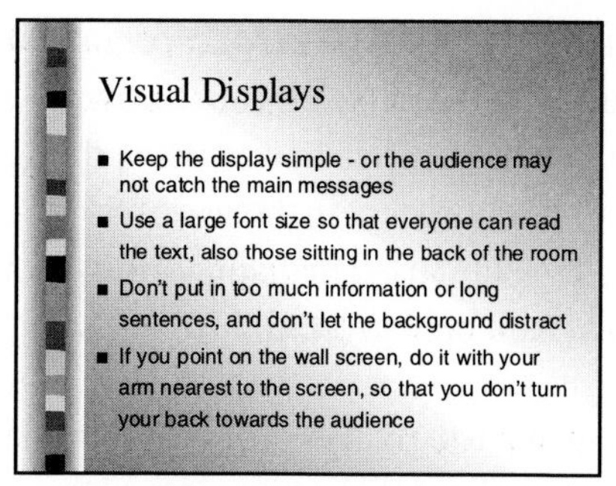

Figure 8:3a. Overloaded with text. Diagonal background disturbing. Presentation template used.

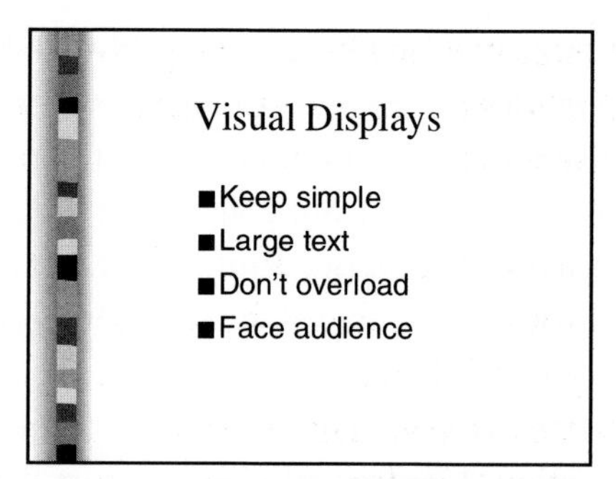

Figure 8:3b. Large brief text (to be supported by the speaker's words). Autolayout adjusted to centre. Light background - room light probably on.

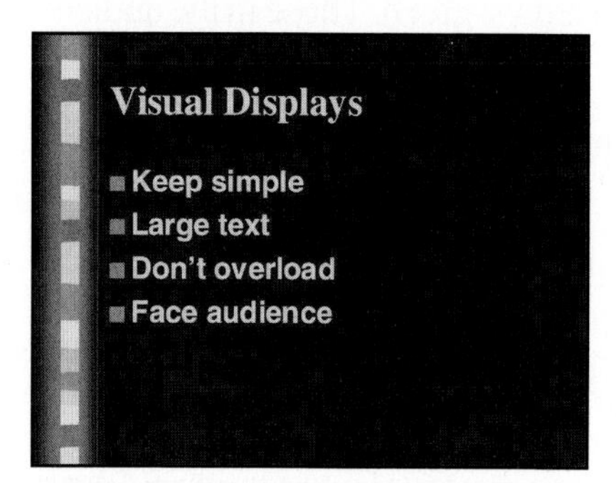

Figure 8:3c. Dark background - room light probably off. Text bold. Autolayout not adjusted to centre.

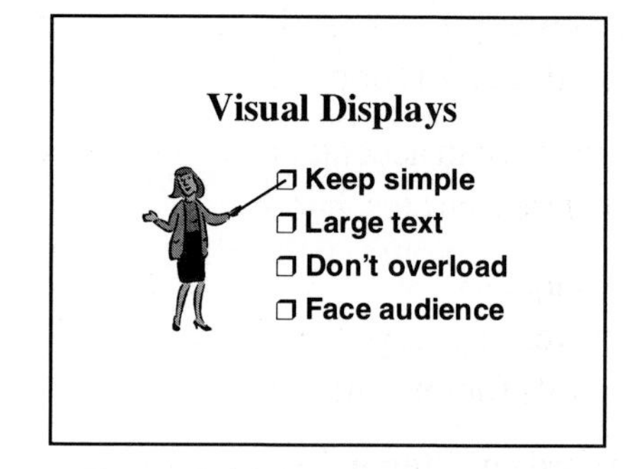

Figure 8:3d. Starting from blank page - no fixed layout. Text bold. Clip art related to content.

Figure 8:3. Examples of visual displays, all depicting the same message.

Use a font size that can be read from a distance

It is an absolute necessity that everyone in the audience can see and read your visuals, including those sitting at the back of the room. The speaker should never need to say: "You probably cannot see this, but ...". If that is what you expect, don't use the visual! The font size might have been sufficient in a small room, or on a large screen, but always be prepared for the fact that the screen may not be large enough. Before making all your visuals, check the readability of a few of them under various conditions if you are not certain of the

Use a large font size

font size needed. When using PowerPoint, for example, you can also look at your slides in "slide sorter view". If they are hard to read, your font may not be large enough.

The standard layouts offered in presentation software are intended to make it easy for you to balance the display and to use a proper font size. Note, however, that you might still need to change the line spacing or to move the text to get balance in the visuals. Therefore, it might be just as good, or even better, to use the option "blank page". Then you will be more active in designing your visuals.

When writing text on visual displays you should think of the following:

- Each display should be simple and easy to grasp quickly. It is usually better to produce two simple displays rather than overloading one.
- Bullets can accentuate the text. Use dots or numbers, or insert symbols like squares, circles, hands, arrows etc.
- A large font size must be used for the text (minimum 20–24 points); headings preferably larger. In the standard layouts the font size is usually around 30 points for text and 40–45 points for headings.
- Words written in lower-case letters (or with an initial capital) are easier to read than words all in capitals.
- Font types without serifs (e.g. Arial or Helvetica) are often said to be more easily read in short text messages, whereas fonts with serifs (e.g. Times New Roman) are easier to read in full paragraphs. If you want it to look less formal, use e.g. Comic Sans.
- Bold text might be seen better, but only if the text is brief.

Show a table or a graph?

Research results are often presented in tables in the written scientific paper. In visual displays for oral presentation, however, tables are often converted to graphs, which usually makes it easier for the audience to grasp the message quickly. If the precise data are important, use a table. But if the major purpose is to show a trend or to make a comparison, a graph can be far more instructive. Below you see the same data illustrated in a table and in a graph (Figure 8:4). Which one of those do you think will best help the audience to catch the message?

Yield (kg) in different years

Year	Variety A	Variety B
1970	75	50
1980	275	150
1990	250	140
2000	375	175

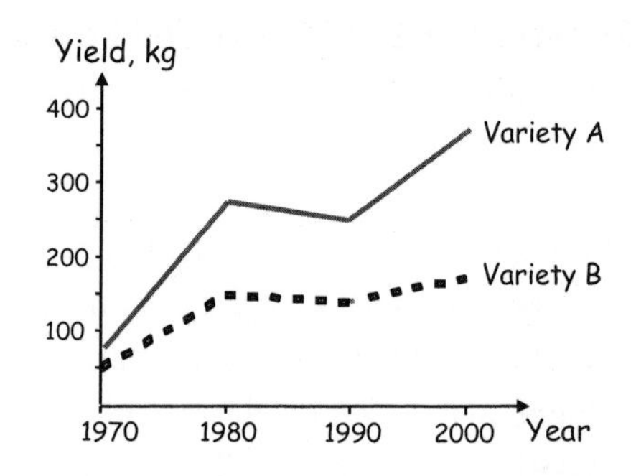

Figure 8:4. The same data illustrated in a table and in a graph (line chart).

As with text displays, both tables and graphs must be simple and not overloaded. The time available for the audience to look at each table or graph will be very limited. Make sure that it is clear what they contain. Abbreviations should be avoided as much as possible, but if you have to use them, make them logical. Help the audience to catch the conclusion(s) from each table/graph. You can, for example, write a summary conclusion in connection with each table or graph shown.

Tables should not be complex. Two or three rows and/or columns are usually enough. The font size must be as large as for text displays. Rounding off the figures adds to clarity. The table visual must not be a non-enlarged copy of the full table in the paper, but could be a part of it.

Making attractive graphs should not be a problem, because there is useful software available for that. Which type of graph to choose depends on what you want to illustrate (see Figure 2:1 in the chapter "Sections of a Scientific Paper"). Use a:

- Line chart (diagram) to show a trend
- Scatter chart to show dispersal or a line fitted to data
- Bar chart to make comparison at specified occasions
- Pie chart to show proportions

Minimise criss-cross reading of your graphs. If possible, write definitions next to each line or pie-segment, instead of having explanatory legends outside the graph.

HOW TO PERFORM THE ORAL PRESENTATION?

A presentation can be done in many ways,
as long as the audience interest stays,
but always be keen,
that what is said is heard,
and what is shown is seen.
Remember the communication ABC,
enjoy the opportunity with an audience to be!

An oral presentation can be performed in different ways. You can choose which type of visuals to use (assuming the equipment needed is available), and also to what extent you use them. You can stand by the lectern, or you can stand away from it and use more of your body language, and even move around a little. There might be a few occasions when you could sit down during a presentation, assuming the audience can still see you, but usually it is much better to stand.

Choose an approach that fits your personality and makes you feel comfortable. But, at all times remember the people in the audience. You need to be in contact with them throughout your presentation.

The audience must hear you and feel involved!

An oral presentation should fulfil the communication ABC, i.e. be Audience-adapted, as well as Accurate, Brief and Clear. This is, however, not the whole story. Some additional basic requirements need to be fulfilled as well:

- Show interest and enthusiasm
- Eye contact with audience
- Speak so that you are heard and understood
- Use only visuals that are clearly seen and support the speech

Your own commitment is fundamental for a successful presentation. Showing interest and enthusiasm should not be difficult. You have a message of value to the audience! Focus on that and on getting your message across. Establish eye contact and make your listeners feel that you are talking to them; you will then also feel their response. Don't let it disturb you if someone in the first rows falls asleep; maybe he/she just had a late night! Concentrate on those who seem interested, but try to look in all directions. Because eye contact is so essential, do what you can to make the organisers keep the room light on, and use visuals that make this possible.

Your voice is important in your presentation. It is essential that everyone in the audience can hear what you say. Remember, when talking at an international meeting, that many people in the audience might be listening to a foreign language. Your speech must be clear. So, speak up and articulate! If using a microphone, talk in conversational tone; the loudspeaker volume can be increased if needed. Practice using a microphone if you are not used to it. Avoid wearing a necklace or other hard items near the microphone. It may easily cause a disturbing noise. Also, dress so that you have a pocket for the battery in case a wireless microphone is used. Otherwise, you may need to hold this in your hand, possibly accompanied by your notes and a pointer.

Speak clearly, but don't shout

Think of your voice speed. If you talk too fast, many people will not perceive what you say, particularly if you don't speak in their mother tongue. On the other hand, if you speak too slowly throughout the presentation, the audience interest may drop. To maintain interest, you can vary your voice level and speed, and now and then also make short pauses.

Is a manuscript needed?

Whether you use a manuscript for the presentation, and what type, depends on what makes you feel comfortable, and how much you have rehearsed. In many situations it is useful to have some kind of manuscript.

The general advice is that you should **not** read your speech from a fully written manuscript. Reading makes it hard to keep eye contact with the audience, and it is usually more difficult to catch what is said if you read aloud, compared with when you talk directly to the audience. If you must read from a manuscript, due to language problems, learn to read whole phrases or sentences, and face the audience with eye contact as much as possible while speaking. It may be easier to look up from your manuscript if you write with a large font size and only put text on the top half of each page.

A manuscript with key words for memory support is often a good solution. If your presentation is based on using visuals, a reduced copy of every visual, supplied with key words to remember what to add in connection with each, can be very helpful. If you plan to stand away from the lectern, it can be handy to have manuscript sheets of a small size (note cards). Make sure, though, that your key words are written in a size that you can easily read.

It might be wise to carefully prepare and write down the introductory words of your speech. It may also be helpful to have the closing words in written form.

Keep to time – rehearse

Rehearsal is often the key to a successful oral presentation, and probably the only way to make sure that you will stay within the time allotted. Considering the labour spent on planning the content and preparing visuals, it would be a waste not to spend the time needed on rehearsing. Rehearse on your own or ask a colleague to listen to you; do whatever feels comfortable. You can also use video or a tape recorder. If you are preparing your first major presentation, try to rehearse in front of a group of friends or colleagues. It is quite possible to rehearse without getting tied to an exact text – as long as you don't read from a written manuscript.

It is very important to keep to the time allotted for your presentation. If you don't, the chairman may close your presentation, otherwise the whole meeting programme might get out of time. So, without proper time planning you may have to finish your presentation before getting to the most essential parts.

The best way to keep control of time is to practice the presentation using a timer. It is recommended that you prepare a talk that will last for 75–80 percent of the allotted time, as it usually takes longer to give the actual presentation, including switching visuals. Furthermore, prepare so that you can be flexible at your presentation; decide which parts to omit if needed. It may happen that the time you actually have at your disposal is shorter than planned.

Feeling nervous is natural

Feeling nervous just before and at the start of a presentation is very natural; even experienced presenters often do. Don't worry about it, adrenalin focuses your mind! Remember that you are the expert on your research; nobody in the room knows more than you do about your work. The best way to cope with nerves is to rehearse; to be well prepared; to know what you are going to say (not word by word, though). Memorise the first sentence of your speech; this will get you off to a good start, and thus ease the rest of your presentation. Once started, you usually don't suffer from nerves any more. Having good visuals, plus some key words for your memory, will keep you on the right course.

It is recommended that you go through your presentation in the morning before your delivery. Visualise that it works well. Inspect the room for your presentation in good time and decide where to stand when you speak. Concentrate for a few minutes before you walk up to give your presentation – and think about the privilege of having an audience interested in listening to your message!

Coping with questions

In most scientific meetings there will be an opportunity for the audience to ask questions to the speakers or to raise issues for discussion, either after each presentation or in the general discussion part of the session. Some of these questions can often be predicted, so you can prepare answers, and maybe have some visual displays with additional information.

Questions show interest, so let the audience understand that you appreciate their questions. Writing down a few key words from each question might help you to remember it. It can be good to briefly repeat a question before you answer. That will let the questioner know if you understood the question, it helps everyone in the audience to hear it, and it gives you some time to think. Occasionally it can be difficult for the speaker to understand or hear a question. You can then ask the questioner to repeat or clarify it.

Your answers to the questions should be short. Don't give a new speech! Give room for more questions instead. If you don't know the answer, say so, or offer to find out. Answer to the whole audience and avoid making it a dialogue between yourself and the questioner.

9

POSTER PRESENTATION

Poster presentations of research results are used increasingly at scientific meetings and on other occasions. Including a poster session as an alternative to oral presentations means that the number of papers/abstracts can be increased; posters can also have several advantages for the presenter.

Research results can be presented effectively in a poster:

- Main messages can be highlighted
- Viewers can study the information at their own pace
- Opportunity for questions and meaningful dialogue between poster presenter and viewers
- The poster might be reused
  - e.g. at the presenter's home institution

Posters require a lot of time for preparation. Meeting organisers thus have a great responsibility in planning a poster session so that both poster presenters and audience get the most out of it. It is important that posters are located where the audience can find them easily, and that time is allotted in the programme for these presentations, which should not coincide with oral presentations in the same scientific area. Poster sessions without the opportunity for meaningful dialogue with the author(s) preclude one of the major benefits of this form of communication.

ATTRACT VIEWERS AND SHOW THE ESSENTIALS

Poster sessions are often filled with a large number of posters, and there is intense competition for audience attention. You need to make the audience curious or interested enough to go closer to your poster, but be aware that you have only a few seconds at your disposal to achieve this. The phrasing of the title and the overall appearance of the poster, therefore, are of utmost importance. To maintain and further arouse interest, your poster needs to have a brief, clear message, so that the audience can quickly grasp the most important points to see how it applies to them. Thereafter they may look for more details, and possibly discuss the topic with you as well.

As with oral presentations, it is important to adapt to the poster audience. Think ahead of their probable questions, both for deciding the content of the poster and for preparing the dialogue part. The poster is often structured like a scientific paper, using headings such as Introduction, Objectives, Methods, Results and Conclusions. Another option could be to use more informal headings, like short statements or questions. Novelty might help make the poster more attractive.

Whatever structure is used, the title of the poster and its number in the meeting programme must be given at the top of the poster, followed by the author names and addresses. In addition, the poster should show why the topic is important, the objectives of the study, the most important results, and the main conclusions and possible implications. The methods used are normally mentioned very briefly, whereas the results (with emphasis on visuals) form the largest part of the poster. You may want to include a lot of information, but remember that the viewers might miss your main messages if you overload the poster!

DESIGNING THE POSTER

Before you start planning the poster, you must know the requirements for height and width as specified by the meeting organisers. Make sure you get it right, so that you don't make a poster in landscape orientation when it should have been in portrait. The meeting organisers may also set rules for how to structure the poster contents, but usually you do it in the

Check poster requirements before designing the poster!

way that you find best. It is important to have some consistency within a poster, but variation between posters may add substantially to a poster session. So use your imagination to make an attractive and informative poster, and remember that the key for a successful poster presentation is simplicity!

Choose layout and content

Before deciding on the poster layout it can be helpful to set up a one-page model in proportional scale, either on paper or on the computer. The content can be arranged in columns running down the poster, or in rows running across it. If the poster is wide, then it might be best to arrange the information in columns, so that the viewer walks along the poster from left to right; especially if many people can be expected to read from the poster at the same time.

The key messages on the poster should be placed where you think the audience will notice them best. Arrange the content in a logical order, maybe starting with the importance of the topic at the top-left, and ending with conclusions at the bottom-right of the poster. If you choose to do it this way, make sure to emphasise the conclusions, and don't hide them at the very bottom of your poster. Another solution might be to place the conclusions centrally on the poster, at eye-height for the audience. There is no single answer as to how to do it. Your poster should be self-explanatory, and the sub-sections could, for example, be numbered to guide the viewers.

Visual displays, such as tables, graphs, photos and other illustrations (e.g. drawings, paintings and clip-art) can make the poster attractive and easy to understand, assuming that they are relevant to the poster topic. Strive to find a balance between the poster text and the visual displays, both with regard to size and proportions of the poster. Some examples of poster layouts are given in Figure 9:1.

A poster needs a unifying background, which separates it from the poster board and neighbouring posters. To achieve this, you can mount the individual elements of the poster on coloured cardboard manually ("multi-part poster"), or you can produce a "single-sheet poster". These alternatives are dealt with in the next section. Remember that the background must not be distracting. It is the message that should be emphasised. If you want to unify groups of data on the poster, then you might use a sub-background colour that harmonises with the main

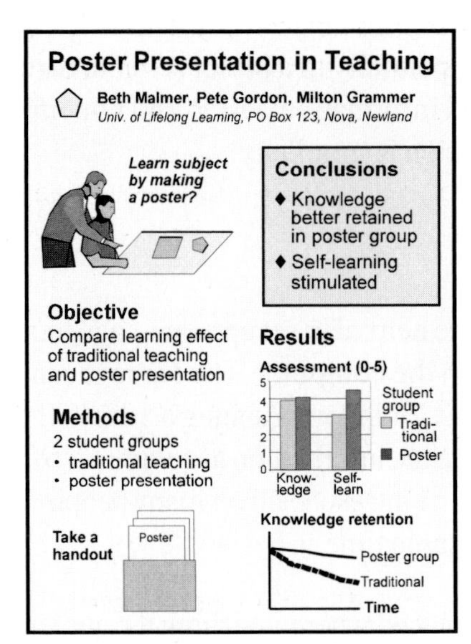

Unifying background. Conclusions at eye-height. Attraction illustration and handout.

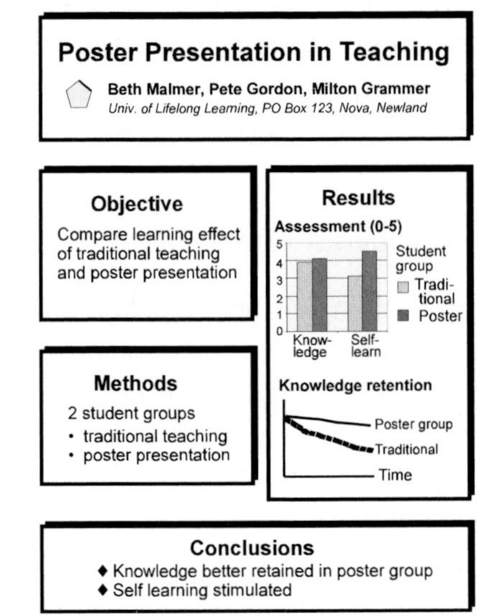

Distinct sections. Poster somewhat "spotted". No unifying background. Conclusions far below eye-height.

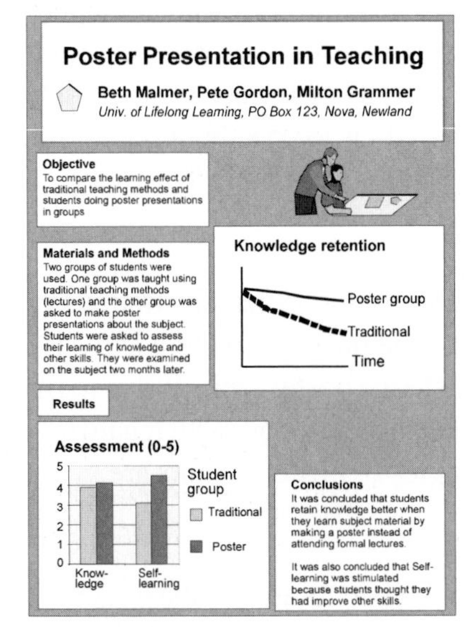

Unifying background. Text on the sheets too small. Visuals larger than needed. Conclusions hidden.

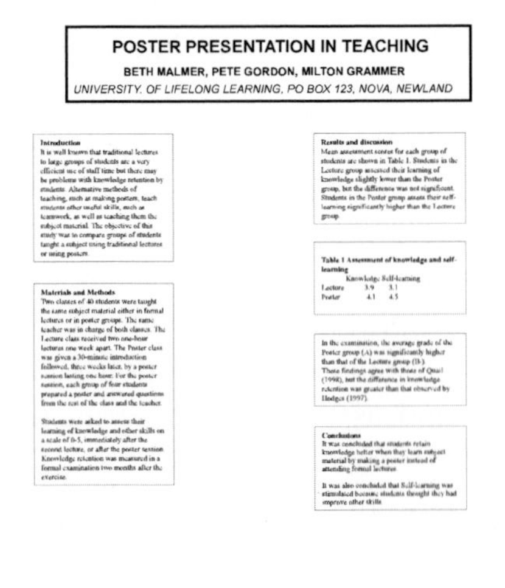

Enlarged manuscript mounted on the posterboard. Should not be accepted!

Figure 9:1. Examples of posters.

background. If you want to accent part of the poster, then you could choose a contrasting colour. Text is usually read best when written on a background that is light (e.g. light beige or grey), but not pure white.

Occasionally you see posters that consist of a number of individual sheets mounted directly on the poster board, often with a dark frame around each and some empty space between the sheets. This usually gives a spotted impression, and we don't recommend this approach.

Each section of the poster should contain just a few messages. You do not need to write complete sentences; so delete most of the words, but leave the meaning. The details can be given in a written paper. For example:

Description of Materials and Methods in printed summary for handout:	**Description of Materials and Methods in the poster:**
Two hundred people, who regularly attended the Heart Clinic at the Royal Free Hospital, were randomly assigned to two equal groups of 100. One group was asked to consume two eggs per day for thirty days and the other group acted as controls. Blood samples were collected from each patient at the start and end of the trial period for cholesterol analysis by the method of Smith (1996). Patients were weighed at the time of blood sampling.	Treatments (30 days duration) A 2 eggs per day (n=100) B Control (n=100) Measurements (days 0 and 30) Plasma cholesterol Body weight

Don't overload the poster! Leave empty space; that is important if the viewer is to catch the content; but make sure that the text and graphics don't appear unrelated. Remember the purpose – to awaken interest and stimulate discussion.

Make the poster

Just like visual displays used in an oral presentation, posters can be created on the computer, either partly or fully. You can use software for word processing, presentation, graphics, drawing and layout. Before making the individual elements of the poster, you should decide how it will be mounted. A poster can be produced in different ways, e.g.:

- *Multi-part poster*

  By "multi-part poster" we mean a poster where the individual elements are produced separately (usually with a computer) and mounted manually on a joint background paper or card. To make the poster stable, the background paper can be pasted onto cardboard or foam-core board. The poster can be given "life" and a deeper dimension if some parts, e.g. illustrations and graphs, or some demonstration materials are attached so that they stick out a little from the background.

  The multi-part poster often needs to be split into segments for transport as carry-on luggage. The final mounting is done at the meeting site, where the segments are put together with broad tape on the back. The poster can be completed in large parts while still at home, but save some text sheets, illustrations etc. to paste at the final mounting, so that you can hide parts of the joints. The final mounting is easier if the segments are taped and folded in pairs before transport.

- *Single-sheet poster*

  If you want your poster to be produced as a single sheet, you can create it digitally on the computer screen by using layout or presentation software, and then print it on a special type of printer to achieve the desired size. Another option is to mount the individual poster elements on background paper, and then make a photographic reproduction. The single sheet poster can be printed on soft paper (and might also be covered with plastic laminate), or it can be printed on cloth, which makes the poster easy to transport. To transport the paper sheet safely, you might need a poster cylinder.

It is not easy to say which type of poster will be best. The single-sheet poster is simple to mount at the meeting site, but the equipment needed to produce it may not always be available, or may be expensive to use. If well done, the multi-part poster might be more "alive" than a single-sheet poster, but be prepared to spend time at the meeting site on the final mounting. The ultimate preference is largely a matter of taste. What you should never do, however, is just to enlarge your written paper to form a poster; this is guaranteed to look dull and unprofessional, and people will not waste time reading it.

When making the poster, it is also important to think of the following:

- Colours will enhance the poster, but too many colours will distract or give a disjointed effect. Title and headings can preferably be written in colour, but the body text is usually easiest to read in black (or dark blue). Colour can be used to highlight, separate or associate information. Think of the background colour when you choose colours for headings. Remember also, that colours on a digitally produced poster may not look the same in print as on the computer screen.
- Bullet points are easier to grasp than text paragraphs. You can use regular bullets or insert various symbols as bullets; the font Monotype Sorts, for example, gives you many options.
- The font used should be easy to read and can be either with serifs (e.g. Times New Roman) or without serifs (e.g. Arial and Comic Sans). Use a font with proportional spacing between characters, rather than one with fixed pitch (e.g. don't use `Courier`). Bold letters in the title and headings may facilitate reading from a distance. Words in lowercase letters (or with an initial Capital) are easier to read than words in all UPPERCASE letters.
- Text size must be large. If the room is crowded it might be difficult for the audience to come very close to the poster. The poster title should be easily read from a distance of 3–5 m and the text from 1.5–2 m. The font size (points) needed for this is about 110–120 for the title, 60–70 for headings and 30–40 for the body text.
- Tables and graphs must be easy to read and to understand (for examples of different types of graphs, see Figure 2:1 in the chapter "Sections of a Scientific Paper"). Use an appropriate font size, limit the amount of information, and enable the viewer to quickly grasp what the table or graph is about. A written conclusion/take-home message directly over or under a table or graph might also help the viewer.
- Clip-art can be useful to illustrate the poster; modify or compose clip-art to fit your purpose if necessary. You might also put an "attention-getter" above or under the poster. The attention-getter could be made in cardboard, for example, or it could be a striking photograph.

- A matt poster surface is usually preferable to a glossy one, because light reflecting from a glossy surface can make your poster impossible to read.

PRESENTING THE POSTER

Don't forget to bring what you might need for the final mounting of the poster, e.g. push pins, glue, spray-adhesive, tape, adhesive dough. To help the audience know who to address for questions and discussion, a photo of the presenter can be fastened near the poster title. You can also attach a small box with your business cards, so that those who want to can easily contact you later. It might also be a good idea to produce a one-page handout and hang copies of that near the poster, so that those who are interested can take them. Such a handout could include the poster title, authors, addresses (also e-mail), a summary of the research, important tables or figures (reduced), and relevant literature. It could also be a reduced copy of the poster.

For the poster session, make sure to arrive on time by your poster, and stay there for the whole session. You might be asked to prepare a 2–5 minute presentation to quickly guide interested viewers through your poster. Remember, though, that the poster should mainly be self-explanatory. Your major role is to be prepared to discuss your topic, respond to questions, and provide additional information if needed. A folder with additional, easily viewed, information can be useful for this purpose. Discussions at a poster session are different from those in an oral presentation session. The poster discussion is more detailed and on a one-to-one basis; it's a dialogue that also gives you a splendid opportunity to establish valuable contacts!

10

TRAINING STUDENTS IN WRITING AND PRESENTATION

Children learn to communicate from birth; they normally learn to speak within three years and are taught to write in school. The development of communication continues through adolescence, so that when students enter university they should have considerable experience in verbal and written communication. Unfortunately most of the emphasis in pre-university education is on telling stories and writing expressive prose or poetry. In scientific subjects, however, more marks are usually awarded for the number of facts recalled than for writing quality. Style is the main thing that distinguishes scientific writing from writing fiction, since science demands precision and accuracy. Many university students are clever enough to understand their work, but cannot communicate their knowledge and ideas effectively. To demonstrate their true potential, students need help with their writing as well as instruction in their chosen subject.

Communication skills are transferable, so if these skills are integrated into a degree programme at an early stage, students will benefit throughout their studies. Communication skills are also beneficial when students seek employment after graduation, even if the students are not pursuing a research career. In a survey of skills required in graduates, both employers and graduates listed communication as the most important skill.[1] There may be some reluctance on the part of lecturers to include communication skills in their courses

Training and practice
- a key to communication skills

[1] CLUES (1998) Personal communication: CTI Centre for Computer Based Learning in Land Use and Environmental Studies, Aberdeen – Skills For Life Project.

because this reduces the time available for teaching their subject. However, the improvement in quality of essays, written reports and class discussions should offset this. With careful planning, students can be taught to communicate subject-specific material so that they learn communication while studying a topic. It is well known that students learn more from doing a task than from being shown or told about the task. Every teacher knows that the best way to learn a subject thoroughly is to explain it to someone else. Therefore, integration of communication skills into the curriculum, using topics that are relevant to their course, should be seen as an effective way of getting students to learn material in more depth.

UNDERGRADUATE AND MASTERS LEVEL

The communication requirements of undergraduate students change and evolve as students progress through their degree programme. These changes often coincide with a reduction in class size as specialisation increases. Initially students have to communicate ideas and facts through essays and practical reports. Oral communication may be limited to informal occasions, such as asking questions in lectures and discussing practicals with a demonstrator or fellow members of a practical team. Later they can be trained to develop presentation skills when class sizes make this more practicable and less daunting for the students.

General study skills

All undergraduates need to learn basic study skills, such as how to manage their time, how to use libraries and computers, how to write essays and practical reports, how to work in groups and how to prepare and present seminars. These skills should be introduced at an early stage of their degree programme and can form part of the induction process. Students can be given a pack that contains advice on important study skills, covering material such as the following:

- Organising your time
- Tutorials (teacher plus 1–6 students)
- Making notes
- Finding information (Literature search, Internet etc.)
- Library skills

- Efficient reading
- Report writing
- Essay writing
- Working in groups
- Giving effective presentations
- Revision and exams

It is even more helpful if a member of staff goes through the information with small groups (10–15) of students. Teachers can emphasise that some of the material covered is not immediately relevant, but it is included at this stage because a) the exact time when individual skills are required varies between subjects and b) it is better to have all material on study and communication skills gathered together in one document. This type of induction course represents an intensive use of staff time, but this is considered to be a valuable use of resources since it saves a lot of individual remedial work later.

As mentioned previously, integration is the key to successful training in communication skills and students learn more effectively by practising the use of these skills. Throughout the degree programme, students can use the skills offered in an induction course in many aspects of their work. It is vital that students are encouraged to improve by giving feedback on their performance (see chapter "Reviewing Papers and Presentations"). All essays, reports and other forms of coursework should be marked to an agreed convention, such as that shown in Table 10:1. In addition, staff must be encouraged to give written feedback on individual pieces of work before returning them to students.

In the later stages of a degree programme, training in communication skills can be more direct and focused on individual students or small groups. Subject areas may differ in their approach, but students can be given a broad training in presentation of scientific ideas through oral and written communication, and posters. This alternative to traditional lectures makes learning more enjoyable and effective for students and can make life more interesting for the lecturer.

Table 10:1. Example of a marking convention for written work submitted by undergraduate and Masters students

Performance	*Marks range (%)*	*Comments*
Outstanding	90 – 100	Outstanding, excellent or very good: • structure, arguments, evidence of critical thinking
Excellent	80 – 89	• understanding, comprehensive and correct knowledge, use of relevant examples • evidence of wider study well beyond set classes.
Very good	70 – 79	
Good	60 – 69	Good structure, some evidence of critical thought, some errors and omissions. Usually few conceptual errors, and some evidence of wider study.
Satisfactory	50 – 59	More errors or omissions, poorly structured with limited argument. Factual information and detail limited; no evidence of wider study. A grasp of the question should be evident.
Barely satisfactory	40 – 49	Poorly structured with little evidence of critical thought or argument. Significant factual error and no evidence of wider study. A basic answer indicating a limited grasp of the question.
Weak	30 – 39	Some indication that the question has been understood, but so much lack of structure and information that a pass cannot be awarded. Extreme omission and factual error.
Extremely weak	20 – 29	Limited evidence that the candidate has understood the question. Extreme superficiality.
Totally inadequate or irrelevant	0 – 19	A response so flawed or superficial that a complete lack of understanding is evident.

Seminars

A seminar delivered by students is probably the most common alternative to lectures. Students work in a group of four or five and prepare a presentation on a topic relevant to the theme of the module. Preparation time may be about six weeks and the group presents their talk to the rest of the class with the teacher present. Input from the teacher depends on the quality of the students. Good students require little guidance or involvement during the presentation. Poorer students may require guidance at the planning stage and assistance on the day, either through acting as an informed member of the audience (asking directed questions) or enhancing the presentation for the benefit of the rest of the class through correcting factual inaccuracies or omissions. Whatever the standard of the presentation, it is vital that the teacher gives constructive feedback on the performance. Feedback should be positive and encouraging, particularly when given publicly. Giving your first presentation is always a traumatic experience. The teacher will usually feel that he or she could have done it better than the students, but the experience of presentation is as important as the subject content and students should not be discouraged by adverse criticism.

Oral presentation of scientific papers

An extension of the seminar could be a Journal Club, which is where students concentrate on a single paper published in a scientific journal. This activity is best performed by students in groups of three or four. Each group selects one paper from a refereed journal that was published in the past three years. The first task is to write a summary of the paper in the students' own words, according to the rules for a scientific meeting, and submit this for assessment. The next task is to prepare and give a presentation of the paper under conference conditions, e.g. ten-minutes presentation plus five minutes for questions. The presentation session should be chaired by a teacher, who ensures that papers run to the time limits given. To ensure adequate questions from the audience, each group of students could shadow one other group by reading the paper that will be presented. Students enjoy this exercise and it helps them to learn about the structure of scientific papers, writing abstracts, sticking to time and dealing with questions, as well as presentation skills.

Writing "popular science"

As discussed in the chapter "Other Types of Scientific Writing", popular science articles need a less formal structure than papers for scientific journals. The ability to interpret written papers for a non-specialised audience is a valuable skill for graduates to use later in their careers.

Students can be trained to summarise scientific papers and rewrite them in a style suitable for the popular press. Students could be given a scientific paper that has recently been published in a refereed journal and asked to rewrite it in the style of a popular science magazine. To make the exercise more valuable, a journalist could be asked to give feedback on the articles and some might actually be published in magazines.

Posters

Another important method for communication of scientific ideas at conferences and in industry is the use of posters. For this exercise, students can work in groups of three or four to research a topic relevant to the module. They can produce an abstract and a poster to illustrate their findings. These are displayed in a poster session, where staff and students view the posters and ask questions of the presenters (see chapter "Poster Presentation").

Dissertations

In some universities, undergraduate students undertake a practical research project in which they perform a piece of scientific research over a fairly long period of time; Masters programmes almost always include a research project. These projects may be linked to the university's research programmes, where experiments are already designed but, ideally, students should have some flexibility in their choice of research area and design of the experiment. The research project normally contributes a substantial proportion of the marks for final degree classification. Each student performs a literature review, conducts an experiment, analyses results and writes a dissertation under the guidance of their supervisor.

In addition, students often present the results of their research to fellow students and staff, either orally or as a poster. This presentation differs from group presentations, since students work individually to present a topic with which they have been intimately concerned.

POSTGRADUATES

The training of postgraduate (PhD) students in communication skills is normally done on a more individual basis, but there is some scope for teaching basic skills to a group of students. Some of the most vital skills to encourage in postgraduate students are planning and organisation. Without planning and organisation, students will not be able to complete their studies on time and will have nothing to communicate.

Postgraduates might run out of time

Introductory courses

All postgraduate students should attend an introductory course on writing and presentation skills. This should include discussion of what makes a good presentation as well as practical sessions, which can be recorded using video equipment so that students can observe their performance.

Informal presentations

Postgraduate students should be encouraged to give informal presentations of their research work to staff, visitors and fellow research students. The emphasis in these meetings is not so much on the actual presentation, but more on the discussion of the work. In this way, students start to feel ownership of their work and get used to thinking about ways to explain their results to other people.

Formal presentations to the Department

Formal presentations are sometimes required as part of postgraduate assessment schemes. During their course of study, postgraduates may present their work to the whole department, both orally and using posters. The first presentation might be six months after registration and might consist of a literature review and outline of the project. After about eighteen months, students should have preliminary results to present and discuss. Later, when the project is nearing completion, students can be expected to give a professional presentation of the whole thesis. Because the audience includes staff and students from different disciplines, clarity of presentation for a lay audience is important.

Formal presentations at scientific meetings

All postgraduate students should expect to make a presentation to a scientific meeting. Because students are representing their university in public, it is essential that they are well trained and give a professional performance.

For oral presentations, preparation should start about two months in advance, when students present an outline to colleagues. The main purpose of this presentation is to confirm the content of the paper and assess the visuals, which should be drafted on overhead sheets. The presentation will probably be modified and repeated after about three weeks. On this occasion, the timing of the paper is studied and the content and visuals finalised. Once the visuals have been made into slides, overheads or a computer presentation, there should be least two more rehearsals to fine tune the presentation and think of likely questions that will be faced. If this sort of rehearsal programme is followed, the conference presentation will be much better.

For poster presentations, drafts should be produced and assessed by colleagues. The final poster can be displayed and students should be questioned on the content.

Writing skills

Apart from the initial training course that should be offered to all new postgraduate students, training in writing skills is almost entirely on an individual basis between student and supervisor. Training usually parallels the development of a student's general scientific ability, so that in the early stages of a PhD programme considerable help is often given, whereas in the later stages the student requires less help with style, leaving more time for discussion of content and development of ideas. The role of the supervisor changes from that of mentor to a partner in research.

An annual progress report may be required as well as a formal presentation. This report does not have the rigid style of a scientific paper, but students can be expected to produce a well-structured document that adequately summarises their progress for the assessors. Students normally discuss drafts of the report with their supervisor before submission.

Research sponsors, including research councils, government bodies and industrial companies, usually require progress reports on an annual basis. Postgraduate students are normally expected to have a high degree of input into these reports, especially in the later stages of their PhD programme.

The ultimate goal of a postgraduate student is to write a thesis that is acceptable to the examiners for the award of their degree. In many countries PhD examinations are based upon a written thesis or monograph of 150–300 pages. In other countries, however, a series of 3–5 journal papers, plus a joint introduction and discussion, are the general form for a thesis. Students are expected to complete the bulk of the writing before the end of their period of supervised study. This requires discipline in the planning stages of the project and in sticking to the plan. If progress reports and reports to sponsors have been diligently written by the student and corrected by the supervisor, much of the thesis writing can be editorial in nature. There is often a tendency for students to keep modifying the thesis so that it is delayed. Such perfectionism must be resisted because it will not be appropriate in later employment. The supervisor should also avoid overcorrecting the thesis. This also introduces delays and the thesis is supposed to be largely the student's own work. If adequate training has been given throughout the PhD programme, students should have developed sufficient skill in writing to produce a thesis with an acceptable style, so that the main focus of the supervisor can be on content.

Although the goal of the student is to produce a thesis, the objective of the institution is to produce scientific papers that can be published in refereed journals. This is the main criterion by which the research quality of most universities is assessed.

Scientific papers are also important for employment and career progression of academic staff and, increasingly, research staff in industry. If the thesis is a monograph, the student should draft one or two papers before graduation for a number of reasons – it is good training for the student, the student should be the most expert person in the world on the research programme, it saves the supervisor time, and the student may not have time to write papers while working in a new job.

In summary:

- Communication skills are important for undergraduate and postgraduate students although their needs are somewhat different.
- Requirements for training change as students progress through their degree programmes and methods move from general group training towards specific individual training.
- A variety of approaches can be used to train students. As well as giving a broader education, this also helps to maintain enthusiasm.
- The use of subject-specific material for training means that staff contact time is not increased and it can also enhance the learning experience.
- Postgraduate students can be trained in groups, but most of their training will be on an individual basis.
- Programme planning and writing practice are essential if postgraduates are to complete their thesis on time and also write papers for publication.

11

REVIEWING PAPERS AND PRESENTATIONS

The process of reviewing other people's written papers is often called "refereeing", to distinguish it from a literature review, but you are doing the same thing – critically appraising the positive and negative aspects of someone's work. Assessment of oral and poster presentations is also discussed, mainly in the context of student presentations.

Whether you are reviewing written material, oral presentations or posters, you should always ask the same basic questions:

- Has the author got his or her message across clearly?
- Is the aim clear (hypothesis or problem)?
- Are facts presented clearly?
- Is the source given where needed?
- Do the facts support the conclusions?
- Are the structure and layout right?
- Is anything missing or superfluous?

When reviewing the work of students and young scientists, be aware of the effects that your comments may have on their self-confidence, particularly if it is their first attempt at scientific writing or public speaking. Always give feedback in a constructive manner; praise the good bits and try to explain alternatives to the bad bits, rather than just saying they are wrong. This approach is also more professional when dealing with the work of more experienced scientists; you do not impress anyone by being totally negative about a paper.

WRITTEN PAPERS

When refereeing or reviewing a written paper for a journal, you need to consider the general points listed above, and you also need to consider whether the science is good enough. The following questions might be asked:

- Does the title of the paper accurately reflect its content?
- Is the abstract complete and does it stand alone?
- Is the subject material relevant to the journal?
- Is the experimental design appropriate for the hypothesis being tested?
- Are the methods appropriate and fully described or referenced?
- Pay particular attention to the statistical analysis – is it correct?
- Are the results presented properly?
- Is material duplicated in text, tables or figures?
- Does the author describe non-significant differences as though they were significant?
- Could alternative conclusions be drawn from these results?
- Does this paper make a significant contribution to the literature?

The journal may provide a checklist for a referee that contains additional points to consider. If the paper contains calculations, check some of them against the data provided; if the results do not agree, ask for all the data to be checked by the author. Check whether all the references are listed in the reference list and that every paper in the reference list is mentioned in the paper; this should be the author's job, or the editor's, but it is surprising how many papers are incomplete.

You will be asked to write a report about the paper and should give a clear recommendation as to whether the paper should be accepted, returned to the authors for correction, or rejected. Make sure your report is constructive and gives sufficient detail about each criticism, so that the author can clearly understand what you require. As a reviewer, you usually have to do considerably more work when recommending to reject a paper than when recommending to accept one. Try to start with some positive remarks; that will make it easier for the author to take the critical remarks. Outline any general criticisms of the paper as a whole, giving references to support your arguments if necessary. The paper should have page and line numbers, so refer to the relevant part of the paper for each point. Move on to minor queries or mistakes, listing these in the order they appear in the paper. Finally, round off with a general conclusion about the value of the paper, i.e. is it really worth reanalysing the data, or are there fundamental flaws in the design? Remember that you have to uphold the standards of the journal.

Return the manuscript and your report promptly to the editor. Make your decision clear in a covering letter. Be prepared to look at the manuscript again after the author has attended to corrections. Your arguments may be challenged, e.g. if the author has read the section on "Editor's and Referees' Reports" in the chapter on "Getting a Paper into Print". Admit when you have made a mistake, but point out that the relevant section should be written more clearly if you, as an expert, found it ambiguous.

STUDENTS' WRITTEN WORK

The same principles apply to students' written work as to journal papers, but your feedback should form part of the learning process. If you are a student reviewing another's work, consider the sort of information that you would need to help improve your written work; if teachers give insufficient feedback, ask them to expand. Always mark to an agreed standard (see chapter on "Training Students in Writing and Presentation").

Remember to give constructive criticism AND praise! Focus mainly on general remarks in verbal feedback, but write detailed notes on the manuscript that you return to the student. With students' work, your remarks might be about style as much as content, whereas with a journal paper you concentrate mostly on content. Remember that marks on their own tell students nothing about where they went wrong, so always let them know how a text or presentation could be improved.

Give praise

ASSESSMENT OF PRESENTATIONS

Exercises in communication can form part of the students' overall assessment, as well as providing training for them. Writing skills are obviously part of assessment anyway (essays, practical reports, examination scripts), but oral and poster presentations can also be assessed using appropriate criteria, as follows:

- Quality and content of the abstract
- Design and content of the visual aids/poster
- Structure of the presentation
- Clarity of delivery
- Use of visual aids
- Evidence of being prepared (not reading a script)
- Enthusiasm and contact with the audience
- Timekeeping
- Handling of questions

It is important that both students and staff are aware of the criteria used. Students often perceive the delivery to be the main part of an oral presentation. It is acknowledged by assessors that nervousness can ruin a delivery, even by very able students. Everyone suffers from nerves when performing in public and the ability to overcome this is only attained with practice and encouragement. Fears can be allayed if the importance of ALL criteria are stressed; students should be awarded a substantial proportion of their marks for the quality of their handout and visual aids, which has little to do with the actual delivery (although there is usually a strong correlation between all aspects of the presentation).

Problems of assessment may arise where students work in groups. Teamwork itself is a useful skill to develop, and individual presentation and assessment are often impractical with larger class sizes. In all teams there are leaders and followers; sometimes groups of students are aggrieved if one member of the team

Develop teamwork skills

does little work towards the presentation. One way around this problem is to award the team a mark and to ask them to negotiate individual marks according to the amount of effort put in.

A member of staff may not always be the appropriate person to assess a presentation. They usually have preconceived ideas about the content, greater knowledge of the subject material and, with the benefit of experience, may have presented it differently. An alternative is to use a scheme of peer assessment whereby each student in the audience marks the presentations according to the set criteria. This system has further advantages in that it makes the audience more aware of the elements of a good presentation and the audience can give an indication of how much they have learnt during the presentation from first-hand experience.

FURTHER READING

The following list of books and web-sites might be useful when writing and presenting a scientific paper for publication. The list, however, is not exhaustive.

BOOKS

Alley, M. 1996. The craft of scientific writing. 3rd ed. Springer-Verlag, New York.

Anholt, R.H. 1994. Dazzle 'em with style - the art of oral scientific presentation. W.H. Freeman and Company, New York.

Booth, V. 1993. Communicating in science: Writing a scientific paper and speaking at scientific meetings, 2nd ed. Cambridge University Press, Cambridge.

Briscoe, M.H. 1996. Preparing scientific illustrations: a guide to better posters, presentations and publications. 2nd ed. Springer-Verlag, New York.

Davis, M. 1996. Scientific papers and presentations. Academic Press, San Diego.

Day, R.A. 1996. Scientific English: a guide for scientists and other professionals. 2nd ed. Oryx Press, Phoenix.

Day, R.A. 1998. How to write and publish a scientific paper. 5th ed. Oryx Press, Phoenix and Cambridge University Press, Cambridge.

Huth, E.J. 1994. Scientific style and format: The CBE manual for authors, editors and publishers. 6th ed. Cambridge University Press, Cambridge.

Lindsay, D.R. 1995. A guide to scientific writing. 2nd ed. Addison Wesley Longman, Melbourne.

Locker, K.O. 1999. Business and administrative communication. 5th ed. Irwin Professional Publishers.

Matthews, J.R., Bowen, J.M. & Matthews, R.W. 1996. Successful scientific writing: a step-by-step guide for biomedical sciences. Cambridge University Press, Cambridge.

McMillan, V.E. 1997. Writing papers in the biological sciences. 2nd ed. Bedford Books, Boston.

Meadows, A.J. 1998. Communicating research. Academic Press, San Diego.

O'Connor, M. 1998. Writing successfully in science. E & FN Spon, London.

Penrose, A.M. & Katz, S.B. 1998. Writing in the sciences: exploring conventions of scientific discourse. St Martin's Press, New York.

Strunk, W. and White, E.B. 2000. The Elements of Style. 4th ed. Allyn & Bacon, Needham Heights, MA.

Sullivan, R.L. 1996. Technical presentation workbook - winning strategies for effective public speaking. ASME Press, New York.

Swales, J.M. & Feak, C.B. 1994. Academic writing for graduate students: a course for nonnative speakers of English. The University of Michigan Press, Michigan.

Yang, J.T., Yang, J.N. & Yant, J.T. 1996. An outline of scientific writing: for researchers with English as a foreign language. World Scientific Publishing, Singapore.

Zinsser, W.K. 1988. On Writing Well: The classic guide to writing nonfiction. 6th ed. Harper Reference, New York.

WEB-SITES (All accessed July 2000)

General search engines

The Internet Public Library (collection of search engines)
http://www.ipl.org/ref/websearching.html

Web Ferret (Combined search engines)
www.ferretsoft.com

Writing

Online Writing Lab. Purdue University, USA.
http://owl.english.purdue.edu

Instructions to Authors in the Health Sciences. Raymond H. Mulford Library, Medical College of Ohio. (contains links to journals in the health and life sciences)
http://www.mco.edu/lib/instr/libinsta.html

Ingenta (Free literature searching of academic and professional journals).
http://www.ingenta.com

Oral and Poster Presentation

British Society of Animal Science (BSAS). Presentation Guidelines (preparation of scientific posters, slides and PowerPoint presentations).
http://www.bsas.org.uk

Radel, J. 1999 Effective presentations. University of Kansas Medical Center
http://www.kumc.edu/SAH/OTEd/jradel/effective.html

Mandoli, D.F. 1996. How to make a great poster.
http://www.aspp.org/education/poster.htm

Fisher, B.A. & Zigmond, J. 1999. Attending professional meetings successfully. University of Pittsburgh, USA.
http://www.edc.gsph.pitt.edu/survival

Clip-art

Microsoft Clip-art Gallery (clip-art that can be downloaded by licensed Microsoft users)
http://cgl.microsoft.com/clipgallerylive/

Clip-art for licensed users of Corel products
http://www.designer.com/freebies/clipart.htm

Other clip-art collections:
http://www.clipart.com/
http://www.clip-art-center.com/

INDEX